access to geography

Hodder & Stoughton

A MEMBER OF THE HODDER HEADLINE GROUP

Acknowledgements

Every effort has been made to trace and acknowledge ownership of copyright. The publishers will be glad to make suitable arrangements with any copyright holders whom it has not been possible to contact.

Dedication

To Angela, Rosie, Patrick and Bethany

Orders: please contact Bookpoint Ltd, 130 Milton Park, Abindgon, Oxon OX14 4SB. Telephone: (44) 01235 827720. Fax: (44) 01235 400454. Lines are open from 9.00–6.00, Monday to Saturday, with a 24 hour message answering service.

You can also order through our website www.hodderheadline.co.uk

British Library Cataloguing in Publication Data
A catalogue record for this title is available from the British Library

ISBN 0 340 800275

First Published 2002
Impression number   10 9 8 7 6 5 4 3 2 1
Year   2007 2006 2005 2004 2003 2002

Copyright © 2002 Garrett Nagle

Cover photo shows floods in Dhaka; reproduced courtesy of © Still Pictures/Shehzad Noorani
Typeset by Fakenham Photosetting Limited, Fakenham, Norfolk.
Printed in Great Britain for Hodder & Stoughton Education, a division of Hodder Headline Plc, 338 Euston Road, London NW1 3BH by Bath Press Ltd, Bath.

Contents

1 Climate and weather

KEY WORDS

Air mass A large body of air with relatively similar temperature and humidity characteristics.

Albedo The reflectivity of a surface.

Anticyclone A high pressure system.

Atmosphere The mixture of gases, predominantly nitrogen, oxygen, argon, carbon dioxide and water vapour, that surrounds the Earth.

Climate The average weather conditions of a place or an area over a period of over thirty years.

Cyclone An atmospheric low-pressure system that gives rise to roughly circular, inward-spiraling wind motion.

Front The boundary between a warm air mass and a cold air mass, resulting in depressional or cyclonic rainfall.

Meteorology The study of the Earth's atmosphere and weather processes.

Precipitation All forms of rainfall, snow, frost, hail and dew. It is the conversion and transfer of moisture in the atmosphere to the land.

Troposphere One of the four layers of the atmosphere, which extends from the surface of the Earth to an altitude of 10–16 km. Climate and weather take place in the troposphere.

Weather The state of the atmosphere at a given time and place.

Climate and weather are fundamental to human society. The links between people, weather and climate are very strong. At one extreme are natural hazards which take many lives, such as floods and drought. At the other extreme, the supply of water maintains life, yet in many urban areas the supply is increasingly uncertain. In this chapter we examine the main patterns of global climate in terms of temperature, rainfall and wind systems. We also look at climatic graphs, and how they can provide us with much useful information.

1 Climate and weather

The term climate refers to the state of the atmosphere over a period of not less than 30 years. It includes variables such as temperature, rainfall, winds, humidity, cloud cover and pressure. This is the total experience of weather at a place over a length of time. It refers to not just the averages of these variables but to the extremes as well. The use of 30 years of records is considered adequate, and so many climate statistics are based on the period 1931–60. However, there are a number of arguments against using a 30 year set of data. For example:

- the database is too short
- the years 1931–60 are not representative
- it is impossible to create a 50 year maximum, or 100 year events from 30 years of records.

In contrast to climate, weather refers to the state of the atmosphere at any particular moment in time. However, we usually look at the weather over a period of between a few days and a week. Climate and weather are affected by a number of factors such as atmospheric composition, latitude, altitude, distance from the sea, prevailing winds, aspect, cloud cover and, increasingly, human activities.

a) Climate data

The use of climate data does have weaknesses. For example, the assumption of climatic consistency has been used extensively by planners. This assumes that 30 years of statistics will provide enough data for modelling the future climate. But if climates change, as we know now that they do, 30 years of records from a cold period, or a wet period, will not provide much help as the climate warms, or dries, or becomes more extreme.

Climate evidence from the twentieth century shows a global warming until about 1950, then a gradual cooling, followed by an increase in extreme events. After many decades in which there was limited research into climatic change, the late twentieth century and early twenty-first century have witnessed a huge volume of research. The years 1931–60, on which much of the data are based, were among the warmest and wettest years (until 1960) of the twentieth century. Between 1900 and 1940 average global temperatures were rising, especially in the Arctic. However, extreme conditions were not as frequent. For example, winters in Europe were generally quite warm, and there was a lack of severe conditions. Annual variability of temperature also decreased during this time period. Rainfall increased in drier areas, although there was a pronounced rain-shadow effect in the lee of mountain ranges (due to an increase in westerly winds), and the monsoon became more regular. The patterns are complex.

2 Global temperature patterns

The main source of energy available to the planet is the sun. In some places there are important local sources of heat, for example geo-

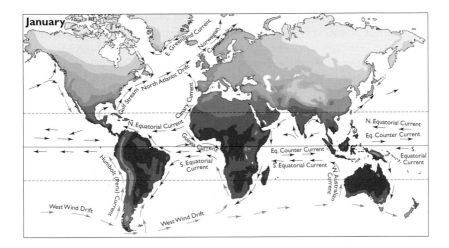

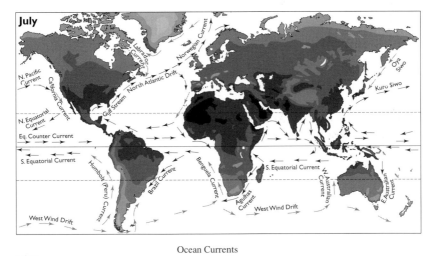

Ocean Currents

→ ⇀ cold

→ ⇀ warm

32
24
16
8
0
−8
−16
−24
actual temperatures °C

Figure 1 Global temperature patterns (January and July)

thermal energy in Iceland and human-related sources in urban-industrial zones. However, solar heating is the main cause of global temperature patterns and also global wind patterns. Although the net radiation (the difference between the incoming radiation and the outgoing radiation) for the planet as a whole is zero, this is not the case in all latitudes. In polar areas the low angle of the sun and the greater *albedo* of snow and ice means that the input of solar radiation is much smaller than in equatorial areas.

There are important large-scale east–west temperature zones (Figure 1). For example, in January the highest temperatures over land (above 30°C) are found in Australia and southern Africa. By contrast, the coldest temperatures (less then −40°C) are found over parts of Siberia, Greenland and the Canadian Arctic. In general there is a decline in temperature northwards from the Tropic of Capricorn, although there are important anomalies, such as the effect of the Andes in South America, and the effect of the cold ocean current off the coast of Namibia. In contrast, in July the maximum temperatures are found over the Sahara, northern India, parts of southern USA and Mexico. Areas in the southern hemisphere are cooler than in January.

3 Air motion

The basic cause of air motion is the unequal heating of Earth's surface. Variable heating of the Earth causes variations in pressure and this in turn sets the air in motion. Most of the energy received by the Earth is in the tropical areas whereas there is a net loss of energy from the polar areas. The major equalising factor is the transfer of heat by air movement.

a) Pressure variations

Pressure is measured in millibars (Mb) and is represented by *isobars*, lines of equal pressure. On maps pressure is adjusted to mean sea level (MSL). MSL pressure is 1013Mb, although the mean range is from 1060Mb in the Siberian winter high pressure system to 940Mb (although some intense low pressure storms may be much lower). The trend of pressure change is of more importance than the actual reading itself. Decline in pressure indicates poorer weather and rising pressure indicates better weather.

b) Surface pressure belts

Sea level pressure conditions show marked differences between the hemispheres. In the northern hemisphere there are greater seasonal contrasts whereas in the southern hemisphere much less extreme conditions exist (Figure 2). The differences are largely related to

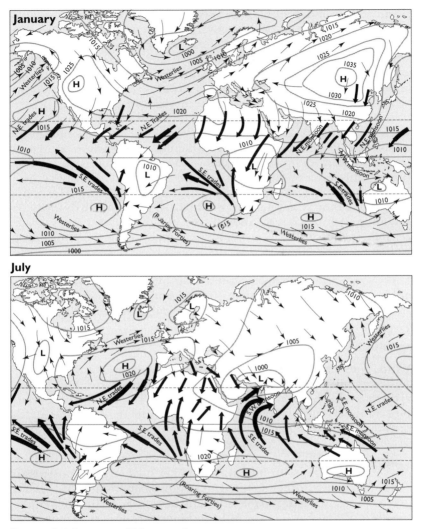

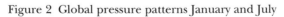

Pressure and winds	H	High pressure cell
Pressure reduced to sea level	L	Low pressure cell

Prevailing Winds
The more regular ('constant') the direction of the wind

Figure 2 Global pressure patterns January and July

unequal distribution of land and sea, because ocean areas are much more equable in terms of temperature and pressure variations.

The most permanent features are the subtropical high pressure belts, especially over ocean areas. In the southern hemisphere these are almost continuous at about 30° latitude, although in summer over South Africa and Australia they can be broken. Generally pressure remains constant at 1026Mb. In the northern hemisphere, by contrast, at 30° the belt is much more discontinuous because of the land. High pressure only occurs over the ocean as discrete cells such as the Azores and Pacific highs. Over continental areas such as south-west USA, southern Asia and the Sahara, major fluctuations occur; high pressure in winter and low pressure in summer because of overheating. Over the equatorial region pressure is generally low. This is due to the zone of maximum isolation which, in July is north of the equator, and in January is just south of the equator.

In temperate latitudes pressure is generally lower than in subtropical areas. The most unique features are the large number of *depressions* (low pressure systems) and *anticyclones* (high pressure systems) which do not show up on a map of mean pressure. In polar areas pressure is relatively high throughout the year, especially over Antarctica, owing to the very low temperatures of the land mass.

c) Global wind patterns

Winds between the tropics converge on a line known as the *inter tropical convergence zone* (ITCZ) or equatorial trough (Figure 3). The convergence zone is a few hundred kilometres wide, into which winds blow inwards and subsequently rise (thereby forming an area of low pressure). The rising air releases vast quantities of latent heat, which in turn stimulates convection.

Latitudinal variations in the ITCZ occur as a result of the movement of the overhead Sun. In June the ITCZ lies further north whereas in December it lies further south. The seasonal variation in the ITCZ is greatest over Asia, owing to the large land mass present. In contrast, over the Atlantic and Pacific Oceans its movement is far less. Winds at the ITCZ are generally light and are called the doldrums, they are broken by occasional strong westerlies, generally in the summer months.

Low latitude winds, those between 10° and 30°, are mostly easterlies (i.e. they flow from the east). These are the trade winds, they blow over 30 per cent of the world's surface. The weather in this zone is predictable – warm, dry mornings and showery afternoons, caused by the continuous evaporation from tropical seas. Showers are heavier and more frequent in the warmer summer season.

Mid-latitude winds are westerly winds. These dominate the area between 35° and 60° of latitude, and account for about a quarter of the world's surface. However, unlike the steady trade winds, these contain rapidly evolving and decaying depressions.

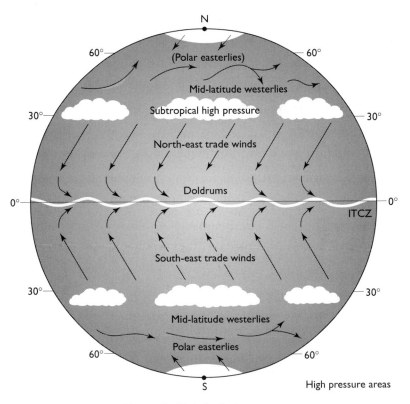

Figure 3 Global wind patterns

Monsoonal winds are the exception to the global pattern. The word monsoon means reverse. The *monsoon* is induced by Asia, the world's largest continent, which causes winds to blow outwards from a high pressure area in winter, but pulls winds into a low pressure region in the summer. The monsoon is therefore influenced by the reversal of land and sea temperatures between Asia and the Pacific during the summer and winter. During the summer the land heats up quickly and may reach over 30°C. In contrast, the sea remains much cooler. This initiates a giant land sea breeze blowing from the cooler sea (high pressure) in summer to the warmer land (low pressure), whereas in winter air flows out of the cold land mass (high pressure) to the warm water (low pressure).

d) Upper winds

Strong westerly waves exist over much of the *tropopause*. These are the thermal winds found just above the polar front where temperature contrast is greatest. The strongest winds are known as the *jet streams.* Jet streams do not flow steadily eastwards but meander in a wave-like pattern. This is known as the *Rossby wave pattern.* The initial deflection of the upper winds could be a mountain range or differences found in coastal areas, for example the relative warmth over a peninsula causes air to rise, and this may deflect the upper air around it. As the upper winds blow *geostrophically* (see page 28) they flow clockwise around the high pressure areas (at high altitude) and anti clockwise around the low pressure areas (at high altitude) (see pages 27–9). Rossby waves include short waves and long planetary waves. The planetary waves can last up to six weeks.

Jet streams are discontinuous and vary, they may even split into distributaries (like a river) and form jet streaks. The polar front jet stream occurs directly above the polar front, and is a thermal wind caused by temperature differences between the air masses involved.

4 Rainfall patterns

Figure 4 shows the pattern of world precipitation. This zonal model shows an abundance of rain in the equatorial zone, moderate to large amounts in the mid-latitudes and relatively low rainfall in the subtropics and at the poles.

A number of zones can be identified. For example, the equatorial zone receives abundant rainfall throughout the year associated with the permanence of the ITCZ. In contrast, the wet and dry tropics experience wet conditions in summer and dry conditions in winter. The summer rains are caused by the ITCZ whereas the winter dry period is due to sub-tropical anticyclones (high pressure systems).

Further away from the equator, tropical semi-arid areas experience a small amount of rain in summer, but very dry conditions in winter. Again these are associated with the equatorial margin of sub-tropical high pressure systems. The arid zones are characterised by permanent dryness and a year-long dominance of sub-tropical high pressure systems. Similarly, sub-tropical high pressure areas experience limited amounts of winter rainfall although occasional mid-latitude depressions may bring some rain.

Mediterranean zones are associated with a long dry summer and a short wet winter caused by sub-tropical high pressure systems in summer and mid-latitude depressions in winter. Polewards, the mid and high latitudes are characterised by depressions and fronts, giving precipitation in all seasons, with maximum precipitation in winter due to an increase in cyclonic activity. In contrast, polar regions have low precipitation due to cold subsiding air, although there may be some depressions in summer.

Numerous modifications to this latitudinal or zonal model occur, for example, heavy orographic precipitation (relief rainfall) is found

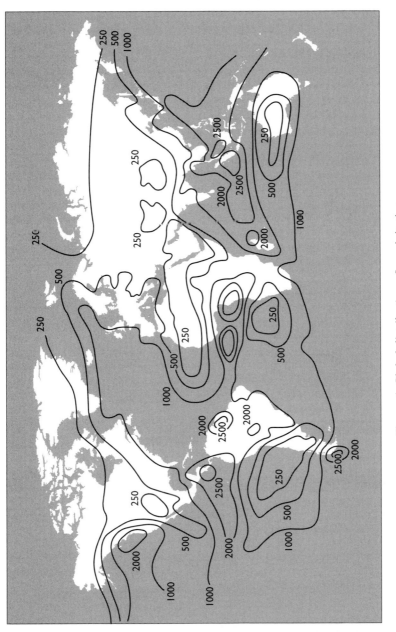

Figure 4 Global distribution of precipitation

along the windward sides of mountain ranges. In the USA the windward side of the Olympic mountains receives 3759mm of rain, the leeward side only 750mm. Similarly, ocean currents affect rainfall. When warm air passes over a cold ocean current, a temperature inversion is formed and fog may form near to the coast. On a larger scale, the monsoon is a result of the seasonal shift of ITCZ and gives rise to intense but strongly seasonal rainfall (see Chapter 3). The mid-latitude cyclonic belt is a zone of mid-latitude depressions moving west to east.

a) Rainfall variability

Rainfall is highly variable (Figure 5). In general, the lower the mean annual precipitation, the higher the variability. Thus, desert areas have a variable precipitation regime. In contrast, humid tropical areas, such as rain forests, have a low variability due to the regular convectional rainfall caused by the location of the inter-tropical convergence zone (ITCZ). Rainfall patterns may also change over time.

5 Reading climate graphs

A *climograph* (or climate graph) is a bar chart representing rainfall and a line graph (or box and whiskers graph) representing temperature over the course of a single year. These simplified graphs tell us a great deal about the seasonal pattern of rainfall and temperature. The **mean monthly average** occurs between the **mean monthly maximum** and the **mean monthly minimum**. The mean monthly maximum is the average of all the maximum temperatures for each day of the month and the mean monthly minimum is an average of all the minimum temperatures recorded for that month. The **absolute maximum** is the highest temperature recorded for that month and the **absolute minimum** is the lowest temperature ever recorded for that month.

Summary

- Climate is the average weather experienced at a place over a period of not less than 30 years.
- Weather is the state of the atmosphere at any given time, generally it refers to a time scale of less than one week.
- Global temperatures show a marked east–west pattern, with temperatures declining towards the poles.
- There are clear seasonal differences in temperature, it is warmer in the northern hemisphere during June–August, and warmer in the southern hemisphere during December–February.
- Air motion is caused by variations in air pressure which are caused by differences in temperature.

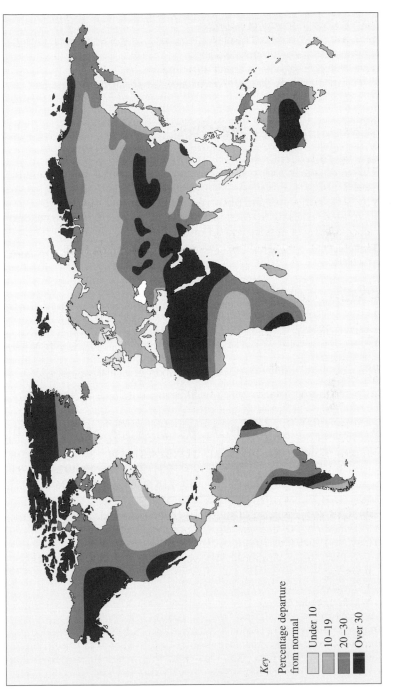

Key

Percentage departure
from normal

Under 10
10–19
20–30
Over 30

Figure 5 Global pattern of rainfall variability

- There are major zones of high and low pressure over the globe – air blows from high to low pressure.
- There are many different rainfall zones, such as the equatorial, savanna, Mediterranean and temperate zones.
- Rainfall is highly variable in space and time.
- Climographs provide data on rainfall and temperature for an area.

Questions

1. Study Figure 4.
   a) Why do equatorial areas experience high rainfall throughout the year?
   b) Why do arid zones and polar regions receive low rainfall?
   c) Explain briefly why the wet and dry tropics (savanna regions) experience a hot wet season, and a hot dry season.
   d) Identify and explain three factors that upset the zonal model of rainfall.

2. Study Figure 1.6.
   a) Define rainfall variability.
   b) How variable is rainfall variability over (i) the British Isles and (ii) the Sahara desert?
   c) How does rainfall variability relate to the map of mean surface pressure (Figure 2).
   d) Suggest reasons for the global distribution of rainfall variability.

3. Plot the following data for St. Lucia, a Caribbean island, on a climograph.

Month	J	F	M	A	M	J	J	A	S	O	N	D
Temperature °C Max	28	28	29	31	31	31	31	31	31	31	29	28
Temperature °C Min	21	21	21	22	23	23	23	23	23	22	22	21
Precipitation (mm)	135	91	97	96	150	218	236	269	252	236	231	198

From the graph you have drawn answer the following questions.
a) Which part(s) of the year experience most rain.
b) Describe annual variations in temperature in St. Lucia.
c) What are the implications of the variations in the annual weather patterns for:
   • farmers
   • water resources
   • tourism?

Advice

These are structured questions and require short answers. The details are in the chapter, the essentials only are provided here.

1. **a)** Equatorial areas receive high amounts of rainfall on account of
   - high temperatures causing evaporation and convection
   - permanent low pressure conditions
   - large quantities of vegetation which maintain water supplies.

   b) In contrast, arid areas are zones of
   - high pressure
   - descending air
   - very limited water resources.

   c) Savannas experience a hot wet season during the summer – associated with rainfall at the ITCZ – a weak low pressure system. During winter, hot dry conditions are associated with dry, descending stable air – high pressure conditions.

   d) Three factors that affect the zonal model of rainfall are
   - the locations of mountains
   - off-shore currents
   - the monsoon.

The rest of the questions should be answered in a similar way. If the question says 'explain' then a few words of explanation are essential for full marks.

2 Global processes

The atmosphere is an **open system** receiving energy from both the Sun and the Earth. Although the latter amount of energy is very small, it has an important local effect, as in the case of urban climates. **Incoming solar radiation** is referred to as **insolation**. This heats up the ground. In turn the Earth radiates energy out to the atmosphere. Global variations in insolation have a major effect on the pattern of temperature, pressure, wind and precipitation.

1 Atmospheric energy budget

The atmosphere constantly receives solar energy, yet until recently the atmosphere was not getting any hotter. Therefore there has been a balance between inputs (insolation) and outputs (radiation).

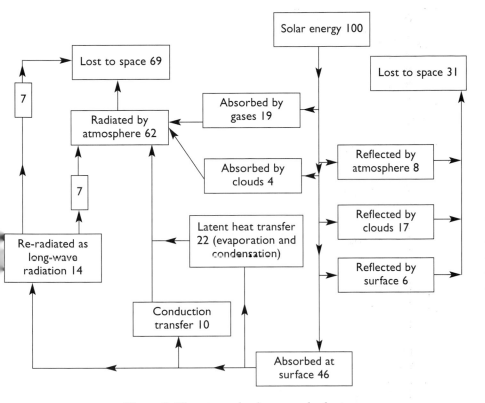

Figure 6 The atmospheric energy budget

Of incoming radiation 19 per cent is **absorbed** by atmospheric gases, especially oxygen and ozone at high altitudes, and CO_2 and water vapour at low altitudes (Figure 6). **Scattering** accounts for a net loss of 6 per cent, and clouds and water droplets reflect 25 per cent. In fact, clouds can reflect up to 80 per cent total insolation. **Reflection** from the Earth's surface (known as the **albedo**) is generally about 7 per cent (Table 1). About 31 per cent of insolation is reflected back to space and a further 19 per cent is absorbed by atmospheric gases. Hence, only about 46 per cent of the insolation at the top of the atmosphere actually gets through to the Earth's surface.

Surface	Albedo (%)
Water (Sun's angle over 40°)	2–4
Water (Sun's angle less than 40°)	6–80
Fresh snow	75–90
Old snow	40–70
Dry sand	35–45
Dark, wet soil	5–15
Dry concrete	17–27
Black road surface	5–10
Grass	20–30
Deciduous forest	10–20
Coniferous forest	5–15
Crops	15–25
Tundra	15–20

Table 1 Albedo values

Energy received by the Earth is radiated at **long wavelengths**. (Very hot bodies such as the Sun emit **short wave radiation**, whereas cold bodies, such as the Earth, emit long wave radiation). Of this 7 per cent is lost to space. Some energy is absorbed by clouds and reradiated back to Earth. Evaporation and condensation account for a loss of heat of 22 per cent.

As a result, the atmosphere is largely heated from below. Most of the incoming short wave radiation is let through, but the out-going long wave radiation is trapped by gases such as water vapour and CO_2. This is known as the greenhouse effect.

2 Explaining variations in temperature

In Chapter 1 we saw that there are important variations in temperature, largely between the equator and the poles.

There are many factors which affect the temperature of a region. These include latitude, altitude, distance from the sea, the nature of nearby ocean currents, dominant winds, aspect, cloud cover, length of day, amount of dust in the atmosphere and human impact.

Changes in temperature between day and night are known as *diurnal* changes. On a calm, cloudless day, minimum temperature is just after dawn since the ground is losing heat all night. After dawn, insolation increases, the temperature of the ground rises, and air temperature rises. This is known as a positive energy budget. Maximum insolation is at midday when the Sun is highest in the sky. However, as it takes a few hours for the ground to heat the air above it, so maximum temperature is sometime around 2 pm. By mid-afternoon convection currents mix warm air with cold air and the temperature begins to drop. After dusk, temperature falls, although the air stays warm for a short while, due to out-going long wave radiation.

a) Latitude

On a global scale latitude is the single most important factor determining temperature. In general, areas closer to the equator are warmer than areas closer to the poles. About 2.4 times as much solar energy is available at the equator compared with the poles. Two main factors affect the temperature:

- the angle of the overhead Sun
- the thickness of the atmosphere.

At the equator, the overhead Sun is high in the sky, hence high intensity insolation is received. By contrast, at the poles, the overhead Sun is low in the sky, hence the quality of energy received is poor. Radiation has more atmosphere to pass through near the poles, due to its low angle of approach. Hence more energy is lost, scattered or reflected than over equatorial areas, making temperatures lower over the poles.

In addition, the albedo is higher in polar regions. This is because snow and ice are very reflective (because of their white colour), and low angle sunlight is easily reflected from water surfaces. However, variations in the length of day and season partly offset the lack of intensity of insolation in polar and arctic regions. The longer the Sun shines the greater the amount of insolation received, so the long summer days in polar regions may overcome the lack of intensity of insolation. (Alternatively, during the long polar nights in winter vast amounts of energy are lost.)

The result is an imbalance in global insolation. There is a positive budget in the tropics (i.e. a heat surplus), and a negative one at the poles (i.e. a heat deficit or shortage). There is a redistribution of heat from areas of surplus (the tropics) to areas of deficit (the poles). To achieve this balance the horizontal transfer of energy from the equator to the poles takes place by winds and ocean currents.

b) Proximity to the sea

The *specific heat capacity* (SHC) is the amount of heat needed to raise the temperature of a body by 1°C. There are important differences between the heating and cooling of water. Land heats and cools more quickly than water. It takes five times as much heat to raise the temperature of water by 2°C as it does to raise land temperatures by 2°C.

Water heats more slowly because:

- it is clear, hence the Sun's rays penetrate to great depth (distributing energy over a wider area)

- tides and currents cause the heat to be further distributed.

Thus a larger volume of water is heated for every unit of energy than land, hence water takes longer to heat up.

Distance from the sea therefore has an important influence on temperature. Water takes up heat and gives it back much more slowly than the land. In winter, in mid latitudes sea air is much warmer than the land air, therefore onshore winds bring heat to the coastal areas. In contrast, during the summer coastal areas remain much cooler than the inland sites. Areas with a coastal influence are termed *maritime* or *oceanic* whereas inland areas are called *continental.*

c) Ocean currents

The effect of ocean currents on temperatures depends on whether the current is cold or warm. Warm currents from equatorial regions raise the temperatures of polar areas they flow into (with the aid of the prevailing westerly winds). The Gulf Stream and North Atlantic Drift transport heat northwards and then eastwards across the North Atlantic; the North Atlantic Drift is the main reason why the British Isles have mild winters and relatively cool summers. Cold currents such as the Labrador Current off the north-east coast of North America may reduce summer temperatures, but only if the wind blows from the sea to the land (see Figure 1 on page 3).

d) Altitude

In general, temperatures decrease with altitude. For every 100m of ascent temperatures drop by about 1°C. So, at an altitude of 1000m the temperature would have fallen by about 10°C. At low altitudes heat escapes from the surface slowly because dense air contains dust and water vapour, which retain heat. In contrast, heat rapidly escapes from high altitudes because the thin air contains little water vapour or dust. If temperatures increase with height then a *temperature inversion* is formed.

e) Winds

The effects of winds on temperature depends upon the main characteristics of the wind. In temperate latitudes *prevailing* (dominant) winds from the land lower the winter temperatures, but raise the summer ones. This is because continental areas are very hot in summer but very cold in winter. Prevailing winds from the sea do the opposite, they lower the summer temperatures and raise the winter ones. This is due to the specific heat capacity of water which means that the sea is cooler than the land in summer but warmer than the land in winter.

f) Cloud cover

Cloud cover decreases the amount of insolation reaching the surface and the amount leaving it. If there is no cloud then incoming short wave radiation and outgoing long wave radiation are at a maximum. For example, rain forest areas with thick cloud cover may experience daytime temperatures of about 30°C and night-time temperatures of about 20°C. Deserts, with little cloud cover, may experience daytime temperatures of 38–40°C, whereas nights drop to around freezing. In Britain during winter, on days where there is much cloud cover, day-time temperatures might only be 2–3° warmer than night-time temperatures. In contrast, when they is a lack of cloud cover, the difference between day and night-time temperatures may be as much as 15°C.

g) Aspect

Aspect refers to the direction a place faces. It is noticeable mostly in temperate latitudes. In the northern hemisphere, south facing slopes (*adret* slopes) are warmer than the north facing slopes (*ubac* slopes). The reverse is true in the southern hemisphere.

h) Length of day and season

Variations in the length of day and seasons are caused by the Earth's revolution and rotation, and have a dramatic impact on the temperature of an area. The Earth **revolves** around the Sun every 365.25 days and it **rotates** around its own axis every 24 hours. The Earth's axis is inclined at an angle of 23.5°, thus the overhead Sun appears to migrate from the Tropic of Cancer (23.5°N) on 21 June to Tropic of Capricorn (23.5°S) on 22 December. The higher the overhead Sun, the greater the amount of heat energy received. Midsummer in the northern hemisphere is June, whereas in the southern hemisphere it is in December. The intermediate positions, 21 March and 23 September, are called the *equinoxes*.

Length of daylight varies with the overhead Sun. On the equinoxes, the Sun is overhead at the equator and all places receive 12 hours day and 12 hours night. On 21 June, the *Summer solstice*, the Sun is overhead at the Tropic of Cancer. Hence the northern hemisphere has its longest days and shortest nights. On 22 December, the *Winter solstice*, the Sun is overhead at the Tropic of Capricorn. Places in the northern hemisphere have their longest nights and shortest days.

i) Human activity

Dust and other impurities such as volcanic fall-out may block insolation from getting to the Earth and keep temperatures low. Deforestation releases carbon, which allows insolation in, but does not allow long wave radiation out, and therefore may lead to an increase in temperature. Aerosols destroy ozone, which prevents harmful ultraviolet light from getting to the Earth's surface.

3 Rain formation and lapse rates

Evaporation is the process by which a liquid is transformed into a gas. It depends on three major factors:

- the supply of **heat** – the hotter the air the more evaporation will take place
- **wind** strength – in windy conditions saturated air is removed, and evaporation can continue whereas under calm conditions the air becomes saturated rapidly but is not removed
- the initial **humidity** of the air – if air is very dry then strong evaporation occurs, whereas if it is saturated then very little occurs.

Most condensation occurs when the temperature drops so that *dew point* (the temperature at which air is saturated) is reached. The decrease in temperature occurs in three main ways:

- *radiation* cooling of the air
- *conduction* cooling when the air rests over a cold surface
- *adiabatic* (expansive) cooling when air rises.

Condensation requires a tiny particle or nucleus onto which the water vapour can condense. In the lower atmosphere these are quite common and include SO_2, NOx, dust and pollution particles. Some of these particles are *hygroscopic* i.e. they attract water, hence condensation may occur when the relative humidity is as low as 80 per cent.

a) Rain formation

For rain to occur, three factors must be satisfied:

- the air must be saturated i.e. it has a relative humidity of 100 per cent
- the air must contain particles of soot, dust, ash, ice, etc.
- the air temperature must be below dew point.

Clouds are tiny droplets suspended in air, whereas rain droplets are much larger. There are two main theories of rain formation, the *Bergeron-Findeisen theory* and the *coalescence theory*. According to the Bergeron-Findeisen theory the formation of rain requires water and ice to exist in clouds at temperatures below 0°C. In a cloud where water droplets and ice droplets coexist, air is oversaturated with water in respect to ice. Ice crystals grow at the expense of water droplets. As they become large enough to overcome atmospheric turbulence, they fall. As they fall, crystals grow to form larger snow flakes, but as they pass into the warm air layers near the ground they melt and become rain. Thus, rain comes from clouds which exist well below freezing level, where coexistence of water and ice is possible.

Other mechanisms for rain formation include:

* condensation on extra large hygroscopic nuclei
* coalescence by sweeping, whereby a falling droplet sweeps up others in its path
* the growth of droplets by electrical attraction.

The most important of these is probably the coalescence theory. When minute droplets of water are condensed from water vapour, they float in atmosphere as clouds. If droplets coalesce they form large droplets which, when heavy enough to overcome gravity, fall as rain.

The Bergeron-Findeisen theory relates mostly to the formation of ice. Rain and drizzle are found when the temperature exceeds 0°C (drizzle has a diameter of <0.5mm). Sleet is partially melted snow and hail is alternate concentric rings of clear and opaque ice, formed by being carried up and down in vertical air currents in large cumulonimbus clouds. Freezing and partial melting may occur several times before the pellet is large enough to escape from the cloud.

b) Types of rainfall

There are three main types of rainfall (Figure 7)

* cyclonic – uplift of air within a low pressure area (warm air rises over cold air). It normally brings moderate intensity rain and may last for a few days
* orographic – a deep layer of moist air is forced to rise over a range of hills or mountains
* convectional heating causes pockets of air to rise, cool and condense to form rain.

c) Fog formation

Fog is caused by clouds occurring at ground level. In fog visibility is less than 1 kilometre, whereas in mist, it is over 1 kilometre. Fog is common in many areas, for example the North Sea coast of Britain in summer, the Grand Banks of Newfoundland and coastal Peru. Fog occurs when condensation of moist air cools below its dew point. The most common types are radiation fog and advection fog (Table 2).

For fog to occur condensation must take place near ground level. Condensation can take place in two major ways:

* air is cooled
* more water is added to the atmosphere.

(a) Convectional

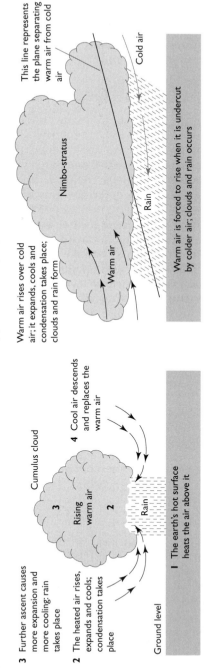

3 Further ascent causes more expansion and more cooling; rain takes place

Cumulus cloud

3
Rising warm air

4 Cool air descends and replaces the warm air

2 The heated air rises, expands and cools; condensation takes place

2

Rain

Ground level

1 The earth's hot surface heats the air above it

(b) Frontal or cyclonic

Warm air rises over cold air; it expands, cools and condensation takes place; clouds and rain form

Nimbo-stratus

Warm air

Rain

Cold air

This line represents the plane separating warm air from cold air

Warm air is forced to rise when it is undercut by colder air; clouds and rain occurs

(c) Relief or orographic

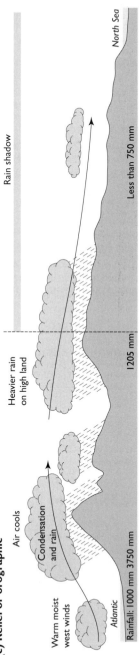

Air cools

Condensation and rain

Warm moist west winds

Atlantic

Heavier rain on high land

Rain shadow

North Sea

Less than 750 mm

1205 mm

Rainfall: 1000 mm 3750 mm

Figure 7 Types of rainfall

Type of fog	Season	Areas affected	Mode of formation	Mode of dispersal
Radiation fog	October to March	Inland areas specially low lying, moist ground	Cooling due to radiation from the ground on clear nights when the wind is light	Dispersed by solar radiation or increased wind
Advection fog (over land)	Winter or spring	Often widespread inland	Cooling of warm air by passage over cold ground	Dispersed by change in air mass or by gradual warming of the ground
(over sea and coastline)	Spring and summer	sea and coasts; may extend a few km	cooling of warm air over cold sea	Dispersed by change in air mass or by gradual warming of the coast
Smoke fog (smog)	Winter	Near industrial areas and large conurbations	Similar to radiation fog	Dispersed by wind increase or by convection

Table 2 Types of fog
(*Source:* Adapted from Nagle G 2000 *Advanced geography*)

The cooling of air is common and caused by orographic, frontal or convectional uplift. In contrast, the addition of moisture to the atmosphere is relatively rare. However, it does occur over warm surfaces such as the Great Lakes in North America or over the Arctic Ocean. Water evaporates from the relatively warm surface and condenses into the cold air above to form fog. Calm, high pressure conditions are required to avoid the saturated air being mixed with drier air above. In addition, contact cooling at a cold ground surface may produce saturation. As warm moist air passes over a cold surface it is chilled, condensation takes place as the temperature of the air is reduced and the air reaches dew point (the temperature at which relative humidity is 100 per cent). When warm air flows over a cold surface *advection fog* is formed. For example, warm air blowing from the North Atlantic

Drift over cold surfaces in Devon and Cornwall will often form fog. Similarly, near the Grand Banks off Newfoundland warm air from the Gulf Stream passes over the waters of the Labrador Current. This is 8–11°C cooler, since it brings with it meltwater from the disintegrating pack-ice further north. This forms dense fog 70–100 days per year. Fog occurs on 40 days per year in the area around the Golden Gate Bridge, San Francisco, because of warm air moving over the cold offshore currents. With light winds, the fog forms close to the water surface, but with stronger turbulence the condensed layer may be uplifted to form a low stratus sheet. *Radiation fog* occurs when the ground loses heat at night by long wave radiation. This occurs during high pressure conditions associated with clear skies.

Fog is a major environmental hazard, airports may be closed for many days and road transport is hazardous and slow. Freezing fog is particularly problematic. Large economic losses result from fog, and the ability to do anything about it is limited. This is because it would require too much energy (and hence cost) to warm up the air or to dry out the air to prevent condensation.

d) Lapse rates

The heating and cooling of air causes changes in the relative humidity and the buoyancy of air, which may lead to condensation and evaporation which in turn can cause cloud formation.

Air is heated from below. When the ground is heated, air that is in contact with it will also be heated. As its temperature rises the air expands. Gases expand when they are warmed and as the air expands its density falls. Consequently the air occupies a much larger volume of space. As air density falls it becomes lighter than the surrounding air and begins to rise. When air is cooled it contracts, its density increases and the air sinks. The heating and cooling of air therefore causes vertical movements within the air. In addition, as air becomes cooler its ability to hold moisture is reduced. For example, if air is cooled to the point where it can no longer hold any more moisture as vapour, condensation takes place and water droplets form. If the air is heated the droplets evaporate and become vapour again. Hence the heating and cooling of air are linked with the processes of evaporation, condensation and the formation of precipitation.

The rising of warm air is a common process. Local heating occurs for a variety of reasons. These include variations in the colour and wetness of a surface. Air above dark coloured and dry surfaces heats up more rapidly than above light or wet surfaces.

The *environmental lapse rate* (ELR) is the vertical change in air temperature away from the ground. Normally temperature falls with height. This is because incoming solar radiation heats the ground through absorption. Some of this energy is transmitted downwards into the soil but most of it is returned as long wave radiation into the atmosphere. Hence heating is greatest closer to the ground surface and declines with altitude. The average rate at which temperature drops with altitude is 6.4°C per 1000m.

One way of looking at instability and stability in the atmosphere is to consider the heating of air above an island. The island converts sunlight to heat much more effectively than the surrounding water (due to the specific heat capacity). Above the island the air will be warmed, density decreases, pressure falls and the air will rise. If air is warmer than the surrounding ELR it will continue to rise, but if it is cooler it will sink. As air moves away from the ground, surface air pressure decreases. As the parcel of air rises it encounters surrounding air of lower density. The pressure confining the parcel of air is reduced and the parcel of air therefore expands. As the air expands, heat is released from it and it becomes cooler. The rate at which air cools with height as a result of this expansion is 9.8°C per 1000m. This is known as the *dry adiabatic lapse rate* (DALR). Adiabatic means there is no heat exchange between the parcel of air and the surrounding air, the air is referred to as dry because it is unsaturated hence condensation does not take place. Air rises at the DALR as long as no condensation takes place. The surrounding air will cool at the ELR. The parcel of air therefore rises until its temperature and its density are equal to that of the ELR. Once the air becomes colder and denser than the ELR it will sink. This is known as *stability*.

In most cases, however, air contains moisture. Even relatively cold air contains some moisture. If the air is cooled enough, condensation will take place. Once this happens the rate of cooling is reduced. In general as air temperature rises, so too does its ability to hold moisture. Consequently warm air can hold much more moisture than cold air. The amount of moisture that is held in the air is expressed in a number of ways. *Absolute humidity* refers to the amount of moisture in grams/m^3 that is held in the air; *relative humidity* expresses this amount as a percentage of the maximum moisture that air of a given temperature can hold. For example, air of 30°C with a relative humidity of 50 per cent may contain about 16 grams of moisture. In contrast, air of 4°C, with the same relative humidity may only contain 2 grams of moisture. Thus relative humidity is temperature dependent. An alternative way of expressing the moisture content of the air is the *saturation vapour pressure curve*. This is the temperature at which condensation occurs. As a rising parcel of air cools it approaches the temperature at which condensation takes place. When the parcel reaches that temperature it becomes saturated and condensation occurs.

As water changes from a vapour to a liquid it releases latent heat. This release of latent heat counteracts the cooling process by warming the air. Thus the air in which condensation has taken place (saturated air) cools more slowly than the DALR. This rate is known as the

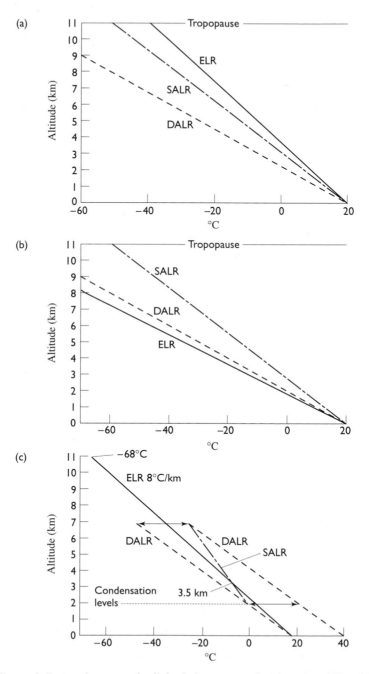

Figure 8 Dry and saturated adiabatic lapse rates showing a) stability; b) instability; and c) conditional instability

saturated adiabatic lapse rate (SALR) and its rate depends on the amount of heat released by condensation. This depends on the temperature of the air and the amount of moisture it contains. Hence, warm air which holds a lot of moisture releases a lot of latent heat. This may cool at a rate of 5°C per 1000m. In contrast cold air, which contains very limited amounts of moisture, will cool at a rate closer to the DALR because less latent heat is released during condensation.

If the ELR is greater than the DALR there is *absolute instability*, i.e. as parcels of air rise they cool at their maximum rate. They remain warmer than the surrounding air and therefore being less dense will continue to rise (Figure 8). If the ELR is less than the SALR stability exists. The rising pocket of air remains colder and denser than the ELR and therefore sinks. If, on the other hand, the ELR is between the DALR and the SALR there is conditional instability, i.e. instability depends on the air reaching its saturation point. Stability and instability are important concepts in terms of cloud formation and rising air. Unstable air rises and produces clouds whereas in stable air uplift and cloud formation are limited, hence precipitation is reduced.

In most cases the ELR lies somewhere between the DALR and the SALR. In this case, i.e. conditional instability, the atmosphere is stable for air which has not reached saturation point (dry air), but is unstable for saturated air. If the air can be forced to reach condensation level, for example by being forced over mountains or by being forced to rise at a front, it becomes unstable and uplift of air is common.

4 Air motion

The basic cause of air motion is the unequal heating of the Earth's surface. Variable heating of the Earth causes variations in pressure and this in turn sets the air in motion. Most of the energy received by the Earth is in the tropical areas, whereas there is a loss of energy from polar areas. The major equalising factor is the transfer of heat by air movement.

a) Factors affecting air movement

The driving force affecting air movement is the *pressure gradient*, i.e. the difference in pressure between any two points. Air blows from high pressure to low pressure. Globally, very high pressure conditions exist over Asia in winter due to the low temperatures. Cold air contracts, leaving room for adjacent air to converge at high altitude, adding to the weight and pressure of the air. In contrast, the air pressure is low over continents in summer. High surface temperatures produce atmospheric expansion and therefore a reduction in air pressure. High pressure dominates at around 25–30° latitude. The highs are centred over the oceans in summer and continents in winter, whichever is cooler.

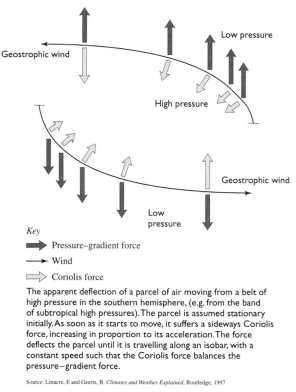

Figure 9 The Coriolis effect

The *Coriolis effect* is the deflection of moving objects caused by the easterly rotation of the Earth (Figure 9). Air flowing from high pressure to low pressure is deflected to the right of its path in the northern hemisphere and to the left of its path in the southern hemisphere.

The balance of forces between the pressure gradient force and the Coriolis force is known as the *geostropic balance* and the resulting wind is known as a geostrophic wind. The **geostrophic wind** in the northern hemisphere blows anticlockwise around the centre of low pressure and clockwise around the centre of high pressure.

The *centrifugal force* is the force experienced when you drive around a corner (objects travelling round a corner are forced outwards). The centrifugal force acts at right angles to the wind, pulling objects outwards.

The drag exerted by the Earth's roughness is also important. *Friction* decreases wind speed, therefore it decreases the Coriolis force, hence winds are more likely to flow towards low pressure.

Figure 2 shows that the main areas of high pressure are at the

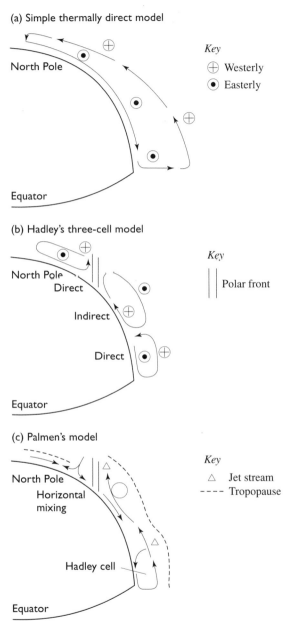

(a) Simple thermally direct model

North Pole

Equator

Key
⊕ Westerly
⊙ Easterly

(b) Hadley's three-cell model

North Pole
Direct
Indirect
Direct

Equator

Key
‖ Polar front

(c) Palmen's model

North Pole
Horizontal mixing
Hadley cell

Equator

Key
△ Jet stream
---- Tropopause

Figure 10 The General Circulation Model

sub-tropics and the poles whereas the main areas of low pressure are located near the equator and in the mid latitudes. Thus we would expect air to blow from the sub-tropics to the equator and to the mid latitudes, and air from the poles to blow to the mid latitudes. Without the Coriolis force this would produce southerly and northerly winds (Figure 9). However, winds are defected to the right of their paths in the northern hemisphere and to the left of their paths in the southern hemisphere. The two dominant forces, the pressure gradient and the Coriolis force, largely account for the surface easterly winds and westerly winds that characterise air motion (Figure 10).

In 1735 George Hadley described the operation of the Hadley Cell, produced by the direct heating of the equator (Figure 10). Air is forced to rise by convection, travels polewards then sinks at the subtropical anticyclone (high pressure belt). Hadley suggested that similar cells might exist in mid latitudes and high latitudes. William Ferrel suggested that Hadley Cells interlink with a mid latitude cell rotating in the reverse direction. These cells in turn rotate the polar cell. The most recent models have refined the basic principles and include air motion in the upper atmosphere, in particular jet streams (very fast thermal winds) (Figure 10).

Summary

The Earth has an energy budget – there is a balance between the amount of heat it receives from the Sun and the amount it radiates back to space.

- Temperature varies globally on account of latitude, distance from the sea, altitude, ocean currents, winds, cloud cover, aspect, season and human activity.
- For rain to fall, air must be cooled below its dew point, this occurs when air blows over a cold surface, or when air is forced to rise at a front, at a mountain range or due to convection.
- Rain droplets may be formed at very low temperatures as cold as −40°C.
- Fog is cloud at ground level, and is formed by the cooling of air.
- Air stability (sinking air) and instability (rising air) are key concepts in the formation of dry and wet conditions.
- Air motion is caused by variations in air pressure. The rotation of the Earth causes air to be deflected from its path.
- The General Circulation Model describes and explains a simplified pattern of air motion for the world, largely moving between high pressure and low pressure.

Questions

1. Study Table 1.
   a) What is meant by the term albedo?
   b) Which surfaces have the highest albedo?
   c) Which surfaces have the lowest albedo?

d) Why is albedo important?
e) Why are light coloured clothes often worn in summer?

2. Table 3 shows the annual temperature patterns for four cities.
   a) Where are the highest temperatures found in January?
   b) How do temperatures change northwards from the Tropic of Capricorn? Are there any exceptions to this pattern?
   c) How do the patterns change in July?
   d) Suggest reasons to explain the changing pattern of temperatures in January and July.
   e) On a world map locate the cities in the table. Using the data show the variations in maximum and minimum temperatures between the four stations.
   f) For each city calculate:
   • the average annual rainfall
   • the average annual temperature.
   g) Describe how the maximum and minimum temperatures vary annually between the four sites.
   h) Using an atlas, account for these differences in terms of:
   • proximity to the sea
   • altitude
   • latitude
   • ocean currents
   • cloud cover.

	Jan	Feb	Mar	Apr	May	June	July	Aug	Sep	Oct	Nov	Dec
Reykjavik (Iceland)												
Daily Max.°C	2	3	5	6	10	13	15	14	12	8	5	4
Daily Min. °C	−3	−3	−1	1	4	7	9	8	6	3	0	−2
Monthly ppt	89	64	62	56	42	42	50	56	67	94	78	79
Timbuktu												
Daily Max.°C	31	35	38	41	43	42	38	35	38	40	37	31
Daily Min. °C	13	16	18	22	26	27	25	24	24	23	18	14
Monthly ppt	0	0	0	1	4	20	54	93	31	3	0	0
Cape Town												
Daily Max.°C	26	26	25	23	20	18	17	18	19	21	24	25
Daily Min. °C	15	15	14	11	9	7	7	7	8	10	13	15
Monthly ppt	10	30	34	44	140	297	325	332	253	114	20	5
Kinshasa												
Daily Max.°C	31	31	32	32	31	28	27	29	30	30	31	31
Daily Min. °C	22	22	22	22	22	19	17	18	20	21	21	22
Monthly ppt	128	142	173	222	129	4	3	3	46	145	246	161

Table 3 Selected data for world sites

Advice

I. (a) Albedo – the reflectivity of the Earth's surface.

    (b) Light-coloured materials such as ice caps – give actual examples and figures.

    (c) Dark-coloured materials such as evergreen trees – give actual examples and figures.

    (d) It influences how much energy a material absorbs (heats up) or reflects (cools down).

    (e) Because they reflect energy, and therefore help keep body temperatures lower.

2. (a) Southern hemisphere and tropical areas.

    (b) They would be expected to decrease towards the equator – rise to the sub-tropical high pressure (STHP) belt, and then decrease again. The stations here fit the model, since Timbuktu is close to the STHP.

    (c) Reversed pattern.

    (d) Due to the Sun now overhead at the Tropic of Cancer in the northern hemisphere.

    (e) Accurate plot is required.

    (f) Statistical manipulation.

    (g) • Reykjavik, Cape Town and Kinshasha are closer to the sea and are at low altitude;
- Reykjavik is high latitude, and others are low latitude;
- Reykjavik receives some warmth from the warm North Atlantic Drift
- Kinshasha has much cloud cover throughout the year, Timbuktu has limited cloud cover throughout the year.

3 Contrasting climates

In this chapter we look at important regional climates. One of the most important regional climates is the monsoon, but there are also important regional climates within areas such as Europe, and even the British Isles. These are the result of important local variations in specific factors, such as altitude, continentality and ocean currents.

1 Cool temperate climates

Cool temperate regions experience a short cold season e.g. Britain, where the low winter temperatures prevent plant growth. These regions are dominated by westerly winds which carry depressions eastwards. Thus there is a contrast between west coast margins (which carry the brunt of the depressional activity) and continental/eastern margin areas which are less affected (Figure 11). In addition, continental areas create high pressure conditions which exclude these depressions. In summer, however, continental areas may become regions of low pressure thus allowing rain bearing winds into the region.

a) Rainfall

Rainfall in western areas is mostly cyclonic, although there is a major orographic impact, and rainfall decreases eastwards, although the

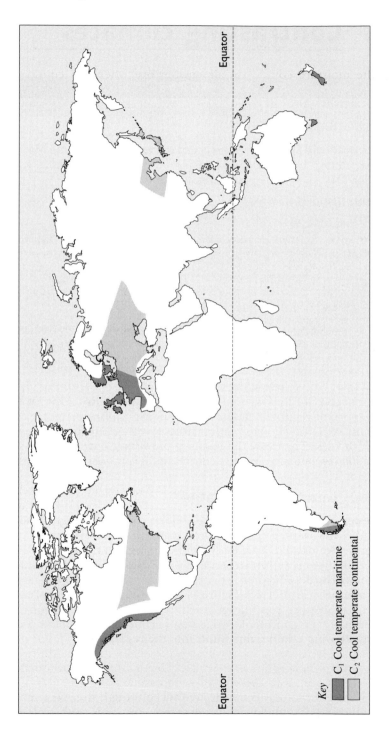

Figure 11 Global distribution of cool temperate climates

pattern is disrupted by mountain barriers. Western areas are characterised by a winter maximum (caused by more depressions) and relief rainfall is greatest in summer due to the warmth of the oceans increasing the moisture content of air passing over them. In continental areas, rainfall is greatest in spring and summer and is due to convectional activity. Also, depressions may reach continental interiors in the summer when low pressure covers the region.

Snow is common in eastern and interior areas in winter. Western areas experience frequent fog, especially in autumn and early winter when the air is moist.

b) Temperature

In western areas the mean annual temperature range may be as low as 8–10°C. Continentality increases eastwards, as does the mean annual temperature range. In parts of the CIS temperatures of over 40°C occur in summer and it gets as low as −30°C in winter! Cool temperate climate can be divided into three main types:

- cool temperate continental type
- cool temperate marine (western type)
- cool temperate continental (monsoonal type)

In general, there is a gradual change from a marine type to a continental type. Nevertheless, the marine type can have exceptionally hot and cold years. The cold winters of 1947 and 1963, and the hot summer of 1976 are excellent examples.

In general, temperatures decline eastwards in winter, due to the lack of oceanic influence. Summer temperatures are the opposite. Although the amount of rain decreases eastwards, proportionally more rain falls in eastern areas in summer. In western areas maximum rainfall occurs in autumn when the sea is still warm and land is cooling.

Northern China is affected by the Asian monsoon. Its climate is characterised by maximum rainfall in summer, and a large temperature range. In winter air blows out of northern Asia bringing cold, dry weather. Occasional depressions may bring limited rainfall. In summer temperatures are high, and the Arctic low pressure sucks in moist warm air.

2 European climates

Europe's complex and fragmented geography provides a variety of climatological and meteorological conditions. In general, there is a north–south divide with southerly areas being warmer than northern locations. In addition, there is a strong east–west factor, with places further east being drier and experiencing greater annual temperature ranges compared with maritime locations in the west. This gives

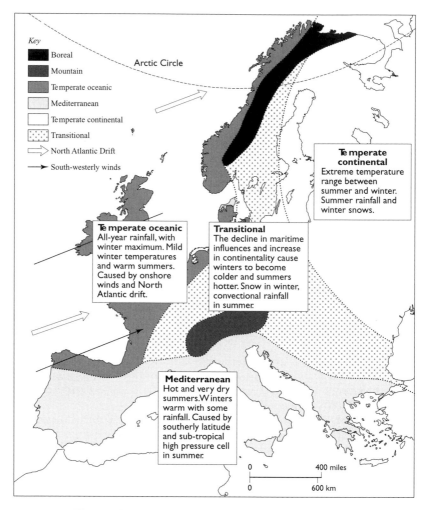

Key

- Boreal
- Mountain
- Temperate oceanic
- Mediterranean
- Temperate continental
- Transitional
- North Atlantic Drift
- South-westerly winds

Arctic Circle

Temperate continental
Extreme temperature range between summer and winter. Summer rainfall and winter snows.

Temperate oceanic
All-year rainfall, with winter maximum. Mild winter temperatures and warm summers. Caused by onshore winds and North Atlantic drift.

Transitional
The decline in maritime influences and increase in continentality cause winters to become colder and summers hotter. Snow in winter, convectional rainfall in summer.

Mediterranean
Hot and very dry summers. Winters warm with some rainfall. Caused by southerly latitude and sub-tropical high pressure cell in summer.

0   400 miles

0   600 km

Figure 12 The fourfold division of European climate

rise to a fourfold division of European climates including a Mediterranean climate in the south, a temperate oceanic climate in the west, a temperate continental climate in the east, and a boreal climate in the north. This pattern is further complicated by the presence of mountain ranges, proximity to the sea (in particular the North Atlantic Drift) and aspect.

The influence of the ocean is most marked in parts of northern Europe. Areas beyond 45° North (Ireland, Britain and Scandinavia) experience positive temperature anomalies (i.e. higher than average temperature for their latitude) in winter. During summer, the

	Jan	Feb	Mar	Apr	May	June	July	Aug	Sep	Oct	Nov	Dec
Shannon												
Daily Max.°C	8	9	11	13	16	19	19	20	17	14	11	9
Daily Min. °C	2	2	4	5	7	10	12	12	10	7	5	3
Monthly precipitation	94	67	56	53	61	57	77	79	86	86	96	117
Number of rain days	15	11	11	11	11	11	14	14	14	14	15	18
London												
Daily Max.°C	6	7	10	13	17	20	22	21	19	14	10	7
Daily Min. °C	2	2	3	5	8	1	13	13	11	8	5	3
Monthly precipitation	52	47	40	48	48	44	56	54	50	52	56	53
Number of rain days	11	9	8	8	8	8	9	9	9	9	10	9
Berlin												
Daily Max.°C	2	3	8	13	19	22	24	23	19	13	7	3
Daily Min. °C	−3	−3	0	4	8	12	14	13	10	6	2	−1
Monthly precipitation	43	40	31	41	46	62	70	68	46	47	46	41
Number of rain days	11	9	8	9	9	9	11	9	8	9	10	9
Moscow												
Daily Max.°C	−7	−6	0	9	17	22	24	22	16	8	0	−5
Daily Min. °C	−14	−13	0	0	6	11	13	12	7	1	−4	−10
Monthly precipitation	31	28	35	35	52	67	74	74	58	51	36	36
Number of rain days	8	7	8	7	8	9	11	11	10	9	9	8

Table 4 Selected data for European sites

ocean has a cooling effect on these places and they experience a negative temperature anomaly (i.e. lower than average temperatures for their latitude). In contrast, continental areas such as Moscow are much hotter during summer but much colder during winter (Table 4).

Europe's location between 36° and 71° North means that there is considerable variation in the amount and intensity of insolation received from the Sun. Annual sunshine varies from about 1000 hours in Iceland to over 3400 in Portugal and south-east Spain.

Most of Europe is dominated by two great pressure systems, the Azores high pressure system and the Icelandic low pressure system. In addition, the Siberian high pressure system influences temperatures during winter, while the Arctic high pressure system and the South Asian low pressure system may be influential during summer. Consequently, most of Europe is characterised by changeable conditions, although southern Europe is more stable during the summer because of the influence of tropical air masses.

These patterns of insolation and air circulation have an important bearing on air temperatures. During January isotherms (lines of equal temperature) run from east to west. The 0°C isotherm divides the warmer south and west of Europe, and the colder centre and north. The effect of ocean currents is most noticeable in this season. During the summer, the ocean has a cooling effect and there is increasing continentality towards the east.

Temperature ranges vary from 8–10°C in Iceland to up to 28°C in parts of the CIS.

Rainfall levels are highest in western areas and over mountainous regions. This is because most of the rain-bearing winds come from the west, as well as the effect of relief on these winds when they reach land. Rainfall varies from between 1000–2000mm in the west of the British Isles to below 500mm in parts of Sweden, southern Spain, Greece and the Baltic States.

Southern Europe has a dry season (spring and summer) which lengthens eastwards and southwards. The rest of Europe receives precipitation throughout the year, with western areas receiving a maximum in autumn and winter. Related to continentality, the proportion of summer rainfall (compared to mean annual rainfall) increases eastwards. In mountainous areas and the interior lowlands of Europe precipitation is most plentiful during spring.

These variations in climate may cause variations in the type and amount of pollution. For example, in areas where precipitation is highest in winter, higher levels of pollutants are washed out of the atmosphere. Areas that are dry in winter may allow a build up of pollutants in the atmosphere. During spring, when plants and animals are growing rapidly, there may be a sudden increase in the level of pollutants in the soil and therefore in the food chain.

3 Climate of the British Isles

The location of the British Isles means that the combined influences of the mid-latitude western winds and the North Atlantic Drift are paramount. The British Isles are surrounded by this comparatively warm oceanic water, the temperature of which varies only slowly from month to month because of the high specific heat capacity of the oceans. Oceanic climates such as Devon and Cornwall have annual fluctuations in temperature of less than 16°C, mild winters, winter maxima of precipitation and moderately warm summers. The east of the British Isles e.g. Gatwick has an annual temperature fluctuation of between 16 and 25°C, mild to moderately cold winters, summer maximum of precipitation and moderately warm summers.

The western and northern parts of the British Isles tend to lie close to the normal path of Atlantic depressions. Consequently their winters tend to be mild and stormy, while the summers, when the depression track is further north and the depressions less deep, are mostly cool and windy.

The lowlands of England have a climate similar to that on the continent (drier with a wider range of temperature than in the north and west). However, the winters are not as severe as those on the continent.

Overall, the south of the British Isles is usually warmer than the north, and the west is wetter than the east. The more extreme weather

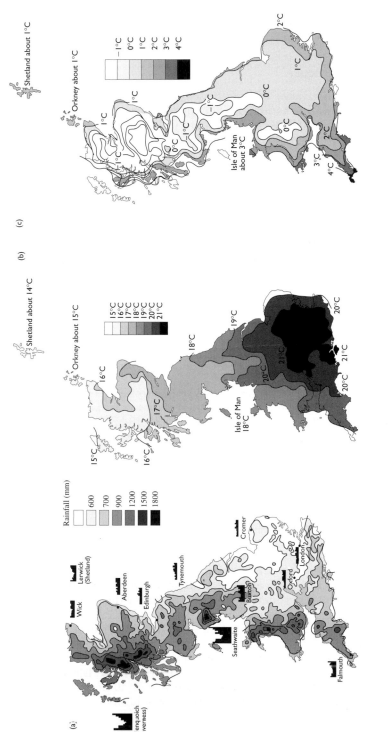

Figure 13 a) Annual rainfall; b) maximum July temperature; and c) minimum January temperature; the British Isles

tends to occur in mountainous regions where it is often cloudy, wet and windy (Figure 13).

a) Temperature

Temperature varies on a daily, seasonal and geographical basis. On a daily basis, the temperature is usually lower at night than by day, with the minimum temperature usually occurring shortly after dawn and the maximum temperature two or three hours after midday.

The temperature varies according to the season because the axis of the Earth is tilted in relation to the Sun. January is, on average, the coldest month and July the warmest. In January, the coldest areas are parts of the Grampian and Tayside regions of Scotland, and the least cold are the extreme south-west of England and the Channel Islands. The main factor determining the distribution of temperature is distance from the coast, particularly the west coast; temperatures are lower inland. In July, the warmest areas are around London and the coolest are in parts of Scotland. Areas near the coast are less warm than inland areas (i.e. the opposite to January), and the temperature decreases from south to north. For both January and July the modifying influence of the sea on coastal regions (keeping temperatures up in winter but down in summer) is mainly felt in a region up to 30km from the coast.

In urban areas the maximum and minimum temperatures tend to be higher than those recorded in rural areas. There are two main reasons for this '*urban heat island*' effect: firstly, the materials used in buildings store heat; and secondly, heat is released as a result of industrial and domestic energy consumption. Some sheltered low-lying areas are more prone to frosts (and have more severe frosts) than the surrounding areas; these are known as 'frost hollows'.

b) Sunshine

Since the length of day varies from winter to summer, the duration of sunshine shows a marked seasonal variation. As a consequence, December is, on average, the month with least sunshine and June is the sunniest. In general, sunshine duration decreases with altitude and increasing latitude, although aspect also plays an important part. For example, south-facing slopes receive more sunshine than those facing north.

Over the year as a whole, the sunniest places are flat areas near the coast. Some sites along the south coast (from the Isle of Wight eastwards) and the Channel Islands record over 40 per cent of the maximum amount possible in a year (1800 hours out of 4000). The

Shetland Islands, on the other hand, only achieve about 24 per cent of the maximum possible sunshine.

c) Rainfall

The mean annual rainfall varies enormously over the British Isles from about 5000mm in parts of the western highlands of Scotland to about 500mm in parts of East Anglia and the Thames Estuary. Overall the wettest areas are in the western half of the country.

The wettest areas occur in the west for two reasons:

* they are nearest to the normal path of rain-bearing depressions
* the most mountainous parts of the British Isles are in the west and, when the moist westerly winds are forced to rise over the mountains, rain is produced.

The south eastern parts of the country have low rainfall because they are further away from the normal track of the depressions. However, much of the Midlands, north-east England and eastern Scotland also have low rainfall because the westerly winds have already dropped much of their water over the mountains in the west. These regions are in a 'rain shadow'.

Although the wettest parts of the British Isles have, on average, ten times as much rain as the driest parts, there is much less difference in the number of rain days (defined as days when 0.2mm or more of rain falls). On average the drier areas have 150 to 200 such days a year, while the wettest areas have just over 200. In most areas, December is the month with the highest number of rain days.

In western areas, the winter half of the year (October to March) tends to receive over half the annual average rainfall. However, in eastern areas there is not such a marked variation, although they generally have more rain in the autumn and less in the spring. In summer, rainfall is often of a showery nature and is normally more intense than the winter rainfall associated with fronts and depressions. The heaviest falls of rain, during summer thunderstorms, can produce rainfall rates of more than 100mm per hour.

d) Snow

The average number of days in a year on which snow is observed to fall increases with latitude and height above mean sea level. Falls of sleet and snow over low-lying areas are normally confined to the period from October to April, although falls do occasionally occur in May and there have been isolated falls in June. However, there are large variations from one year to another in the frequency with which snow falls.

Snow rarely lies on low ground before December or after March. The number of days with snow lying is usually less than the number of days with snow falling because the temperature of the air above the ground generally remains above freezing point and thus melts the lying snow.

e) Wind

The strongest winds are associated with the passage of depressions across or close to the British Isles. As the frequency of depressions is greatest during the winter months, this is when the strongest winds usually occur. The majority of depressions approach the British Isles from the Atlantic, so the windiest areas are the western coasts and hills. Wind speeds decrease away from the coasts, due to the frictional effects of the land. Consequently, western coasts have the highest frequency of gales, especially the western isles of Scotland.

f) Thunder

Thunder can occur in any part of the British Isles at any time of the year, although it is most likely to occur in the summer. East Anglia, the east Midlands and south-east England have the highest occurrence of thunder.

4 The monsoon

The word monsoon is used to describe wind patterns which experience a pronounced seasonal reversal. The most famous monsoon is that of India, but there are also monsoons in east Africa, Arabia, Australia and China. The basic cause is the difference in the heating of land and sea on a continental scale.

In India, two main seasons are identified:

- the north-east monsoon, consisting of (a) the winter season (January and February) and (b) a hot, dry season between March and May.
- the south-west monsoon, consisting of (a) the rainy season of June–September and (b) the post-monsoon season of October–December. Most of India's rainfall occurs in the south-west monsoon.

During the winter season winds generally blow outwards since high pressure is centred over the land (Figure 14a). Nevertheless, parts of southern India and Sri Lanka receive some rain, while parts of northwest India receive rainfall as a result of depressions. These winter rains are important as they allow the growth of cereals during the winter. Mean temperatures in winter range from 26°C in Sri Lanka to about 10°C in the Punjab (these differences are the result of latitude). Northern regions and interior areas, have a much larger tem-

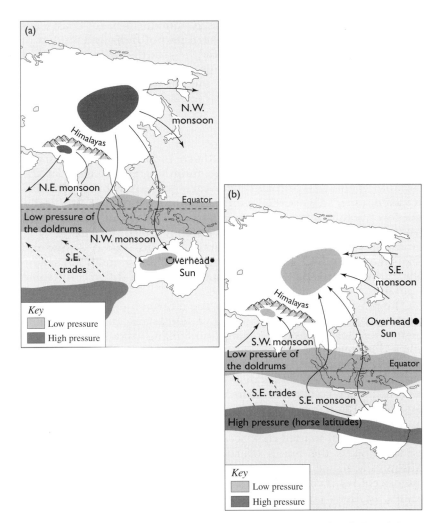

Figure 14 The north-east and south-west monsoons of South East-Asia

perature range compared with coastal areas. In the north, daytime temperatures may reach over 26°C while frosts at night are common.

The hot, dry season occurs between March and May. It gradually spreads northwards throughout India. Daytime temperatures in the north may exceed 49°C, while coastal areas remain hot and humid. Vegetation growth is prevented by these conditions, and many rivers dry up.

In spring the high pressure system over India is gradually replaced by a low pressure system (Figure 14b). As there is low pressure over the equator, there is little regional air circulation. However, there are many storms and dust storms. Increased humidity near the coast leads to rain. Parts of Sri Lanka, the southern part of India and the Bay of Bengal receive rainfall and these allow the growth of rice and tea. For most of India, however, there is continued drought.

The rainy season occurs as the low pressure system intensifies. Once pressure is low enough it allows for air from the equatorial low, and southern hemisphere high to be sucked in, bringing moist air. As it passes over the ocean it picks up more moisture and causes heavy rains when it passes over India.

The south-west monsoon in southern India generally occurs in early June, and by the end of the month it affects most of the country reaching its peak in July or early August (Table 5). Rainfall is varied, especially between windward and leeward sites. Ironically, the low pressure system over north-west India is one of the driest parts of India, as the monsoon has shed its moisture en route. The monsoon weakens after mid September but its retreat is slow and may take up to three months. Temperature and rainfall fall gradually, although the Bay of Bengal and Sri Lanka still receive some rainfall.

The most simple explanation for the monsoon is that it is a giant land–sea breeze. The great heating of the Asian continent and the high mountain barrier of the Himalayas barring winds from the north allow the equatorial rain system to move as far north as 30°N in summer. In summer, central Asia becomes very hot, warm air rises and a centre of low pressure develops. The air over the Indian Ocean and Australia is colder (therefore denser) and sets up an area of high pressure. As air moves from high pressure to low pressure, air is drawn into Asia from over the ocean. This moist air is responsible for the large amount of rainfall that occurs in the summer months. In the winter months the sun is overhead in the southern hemisphere. Australia is heated (Forming an area of low pressure), whereas the intense cold over central Asia and Tibet causes high pressure. Thus in winter air flows outwards from Asia, bringing moist conditions to

	J	F	M	A	M	J	J	A	S	O	N	D	Total
Minicoy (8°N 73°E)	40	35	25	50	205	295	190	180	165	175	150	65	1575
Madras (13°N 85°E)	25	15	20	30	55	55	85	110	105	245	275	125	1245
Mumbai (Bombay) (19°N 73°E)	10	–	–	10	15	505	710	415	300	80	40	–	2085
Chittagong (22°N 92°E)	15	30	50	30	290	525	625	550	330	230	50	25	2870

Table 5 Approximate rainfall figures for selected Indian stations

Australia. The mechanism described here, a giant land–sea breeze is, however, only part of the explanation.

Between December and February the north-east monsoon blows air outwards from Asia. The upper airflow is westerly, and this flow splits into two branches north and south of the Tibetan plateau. The Tibetan plateau – over 4000m in height – is a major source of cold air in winter, especially when it is covered in snow. Air sinking down from the plateau, or sinking beneath the upper westerly winds generally produces cold, dry winds. During March and April the upper airflows change and begin to push further north (in association with changes in the position of the overhead Sun). The more northerly jet stream (upper wind) intensifies and extends across India and China to Japan, the southern branch of the jet stream remains south of Tibet and looses strength.

There are corresponding changes in the weather. Northern India is hot, and dry with squally winds while southern India may receive some rain from warm humid air coming in over the ocean. The southern branch of the jet stream generally breaks down around the end of May, and then shifts north over the Tibetan plateau. It is only when the southern jet stream has reached its summer position, over the Tibetan plateau, that the south-west monsoon arrives. By mid-July the monsoon accounts for over three-quarters of India's rainfall. The temporal and spatial pattern is varied; parts of the north-west attract little rainfall, whereas the Bay of Bengal and the Ganges receive large amounts of summer rainfall. The monsoon rains are highly variable each year, and droughts are not uncommon in India. In autumn, the overhead Sun migrates southwards, as too does the zone of maximum insolation and convection. This leads to a withdrawal of the monsoon winds and rains from the region.

5 Climates in South Africa

South Africa is a country of many climates (Figure 15). The general patterns are:

- rainfall decreases from east to west
- most of the rainfall falls in summer (October to March)
- winter rainfall affects the Cape Town area
- temperatures increase away from the sea
- temperatures are higher in lowland areas, and there is also a general increase towards the north (equator-wards).

Durban has a sub-tropical climate and is warmed by the warm Agulhas current. Summer temperatures are high (20–25°C) and the air is humid. Winters are warm (16–18°C) and frost free. Rainfall is about 1000mm per annum, 70 per cent of it falling during the summer months. Some of this rain is very heavy, and can cause serious soil erosion.

Johannesburg is located on a high plateau, the high veld. Summers are warm (about 20°C) although sometimes the temperatures can be much higher (30°C). Winters are mild (8–10°C) but the nights are very cold, and frosts are common. Temperatures are lower partly as a result of altitude. Up to 85 per cent of the rainfall occurs during the summer with total annual rainfall being just over 800mm. Convection storms (thunder storms) produce heavy rain and can cause local flooding.

The south-west corner of South Africa has a Mediterranean climate. Unlike other parts of the country, it has winter rainfall (April–September). Rainfall is moderate, about 600mm, but can be as high as 3000mm in the mountains. Temperatures range from around

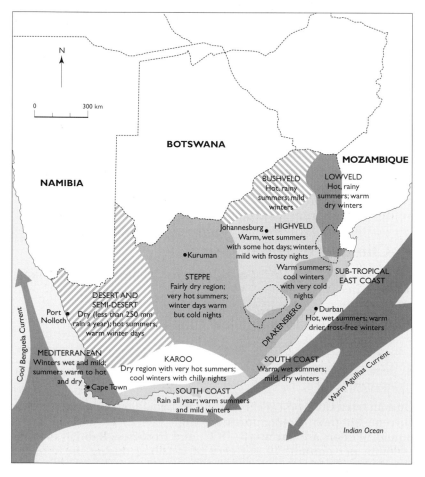

Figure 15 The climate regions of South Africa

20°C in the summer to 15°C in winter. This is due to the moderating effect of the sea.

Relief has an important influence on climate. In general, in South Africa, rainfall decreases from east to west, and over half of the country receives less than 250mm of rain each year. Much of this rain is *relief rainfall*; as the air is forced to rise over a mountain, it cools and condenses, clouds then form and rain occurs. For example, at the base of the Drakensberg Mountains annual rainfall is between 700mm and 1000mm. At the top of the mountain it is almost 2000mm. Similarly, around Cape Town rainfall in the lowland areas is just 400mm, but in the mountains near Stellenbosch, less than 70 km away, it is as high as 3000mm.

Summary

- Climate varies over a relatively small distance, and also seasonally.
- In Europe, areas close to the Atlantic have milder winters and cooler summers than their continental neighbours.
- Climates in Europe become more extreme the further east they are located i.e. they have hotter summers and colder winters than maritime areas.
- Maritime/continental differences can be viewed at a local scale in the case of the British Isles. This is seen in terms of temperature, rainfall and snow cover.
- One of the world's most distinctive climates is the monsoon. This is characterised by a hot dry season and a hot wet season. It is caused by seasonal variations in the distribution of high and low pressure.
- South Africa has a range of climates – Mediterranean, temperate, monsoonal, semi arid as well as mountain.

Questions

1. **a)** Define the terms maritime and continental.
   b) What effect does the North Atlantic Drift have on the climate of the British Isles?
   c) Describe the main characteristics of the climate of Mumbai (see the data in Table 5).
   d) Describe and explain the variations in the climates of the British Isles.

2. **a)** On an outline map of India, locate each of the four stations (in Table 5). If you cannot find them in your atlas, their latitude and longitude are given in the table.
   b) Plot the rainfall figures for Minicoy, Madras, Mumbai and Chittagong using the data in Table 5. (Make sure you use the same scale for each graph). Describe the rainfall patterns you have shown. Suggest reasons for these differences.

3. How and why does temperature and rainfall vary between Cape Town, Durban and Johannesburg?

Advice

I. **(a)** Maritime – mild, equable conditions influenced by ocean currents. Continental – relative extremes of hot and cold caused by distance from the sea.

(b) The North Atlantic Drift warms coastal regions in winters but cools them in summer.

(c) Mumbai – has a hot, wet summer and a hot, dry winter; NB quote figures such as max., min., range, rainfall total and seasonality.

(d) Look for
- north–south variations in temperature (latitude)
- east–west variations (effect of oceans, North Atlantic Drift, continentality)
- east–west differences in rainfall (effect of relief)
- urban–rural contrasts.

As ever, give figures to support your answer.

2. **(a)** Accuracy is essential for the climate plot.

(b) Describe – refer to max., min., range, rainfall total and seasonality, and any exceptions.

Explain – think big –
- northern hemisphere/southern hemisphere
- winter/summer
- highland/lowland
- windward/leeward.

The monsoon is a giant wind system characterised by a seasonal change in direction. It has been likened to a giant land–sea breeze.

In winter (over Asia) it is bitterly cold, and forms a semi permanent high pressure system. Winds blow outwards from high pressure.

In contrast in the southern hemisphere, the overhead sun has warmed the land (low pressure). Winds blow from the Asian high, over the ocean to the Australian low.

In summer, the position is reversed. Australia is relatively cold (high pressure) whereas the overhead sun over Asia produces low pressure. Winds blow from the Australian high pressure, over the moist ocean, producing hot, wet conditions when they are drawn into Asia.

3. Answers should refer to temperature – max., min., range, length of growing season, etc., giving figures from the graph. Similarly, rainfall should include annual totals, seasonality and type of rainfall. Links should be made between temperature and rainfall. Contrasts between Mediterranean (Cape Town), monsoonal (Durban) and tropical wet-dry (Johannesburg) should be made. Remember to refer to

- proximity to the sea
- altitude
- warm/cold ocean currents
- wind systems

for a full explanation.

4 Pressure systems

Figure 2 suggests that mean pressure across the British Isles is approximately 1000Mb. However, this conceals the fact that Britain is characterised by a series of low pressure and high pressure systems. Each type of pressure system produces different types of weather: low pressure systems are associated with rain and wind whereas high pressure systems are associated with clear skies and calm conditions. However, as we will see, no two high pressure systems or no two low pressure systems are the same.

The British Isles lie in the temperate latitude of mainly westerly winds where depressions move across the North Atlantic bringing with them unsettled and windy weather, particularly in winter. Between the depressions there are often small anticyclones which bring fair weather. This sequence of depressions and anticyclones is responsible for Britain's notoriously changeable weather. The air streams associated with the depressions travel across the sea to reach the British Isles. Since the water has a modifying effect (being cooler than the land in summer, but warmer in winter), summers in the British Isles are cooler than those on the continent, but the winters are milder.

1 Air masses

Central to the understanding of low pressure systems are air masses. The idea that northerly winds (i.e. winds from the north) are cold and southerly winds (those from the south) are warm (at least in Britain) is quite common. Similarly, air that has travelled over the sea picks up moisture, while that travelling over the land is relatively dry. These simple concepts help in the understanding of air masses.

An *air mass* is a large body of air often thousands of kilometres wide, with fairly constant temperature and humidity characteristics. Thus polar air masses are cold, tropical ones are warm, those that travel over oceans are damp, and those that have travelled over land are dry.

In some polar and sub-tropical areas air remains in high pressure systems for a long time and begins to take on the characteristics of the underlying surface. Hence air at the poles is cooled and air in the tropics is warmed. The result is a large body of air with little horizontal difference in temperature and moisture content.

Sometimes there is a large outflow of air from the regions where air masses form, and these air masses may approach the British Isles (usually polar air from the north and tropical air from the south). However, on their journey they may be modified by contact with the underlying surface. Air which travels over the sea (maritime air) is moistened, whereas there is little change in moisture content of air which travels over the land (continental air). Air that has been trapped in an anticyclone over the Sahara in June slowly heats up and dries. After a while, the air moves out of the anticyclone and may head for the British Isles. On its way it may collect moisture over the Mediterranean Sea, but the journey over Spain and France has little effect on its properties. The air then arrives as a hot, dry air mass.

Air masses affecting the British Isles can be broadly categorised in terms of their source and their path (Figure 16). This leads to four main types:

- tropical maritime – warm and moist
- tropical continental – warm and dry
- polar maritime – cold and (fairly) moist
- polar continental – cold and dry.

a) Tropical continental

Tropical continental air usually comes with south-easterly or southerly air streams. It originates in North Africa and often travels over the Mediterranean, Spain and France before reaching the British Isles. In summer, easterly winds from central Europe or the Ukraine could be included in this category, as the continent becomes so hot at this time of year. The air picks up some moisture over the Mediterranean, but overall the air tends to be dry and the skies are typically cloudless.

The majority of tropical continental air streams give heat waves in summer. The lack of moisture usually causes the visibility to be good. However, in the air there may be desert dust, fine soil or pollution

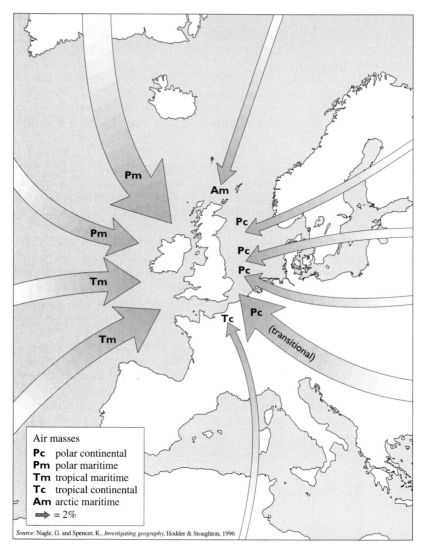

Figure 16 Air masses affecting the British Isles

particles, which can lead to moderate visibility (often described as 'heat haze'). Also, the cloudless sky sometimes looks milky because of pollutants.

b) Polar continental

A polar continental air mass originates in Scandinavia or Russia, and the air mass reaches the British Isles when north-easterly or easterly winds become established. This tends to occur when there is a high pressure area somewhere to the north of the British Isles, often over Scandinavia itself. Polar continental air masses mainly affect the British Isles during the winter.

Temperatures in polar continental air masses are below average in winter. In summer, however, the temperatures tend to be above average. The moisture content is low in these air masses, especially when they take the short sea track in the Channel region. This leads to clouds being well broken, and so the weather is fine and sunny. Air that has crossed the North Sea between Denmark and Scotland has taken a longer sea track. It therefore collects more moisture, and clouds tend to form during its journey over the sea. Consequently, it results in cloudy conditions in eastern areas but further inland there tends to be a mixture of cloud and sunshine. Visibility varies, generally being very good when air comes from Scandinavia, but moderate or poor when the air originates in the industrialised regions of central or eastern Europe. Even in April or May, the North Sea is cold and does little to modify the air mass, apart from adding a little moisture. Southern England is particularly chilled by polar continental air masses. Further north the air streams are less cold and the wind is less strong.

c) Tropical maritime

Tropical maritime air usually approaches the British Isles from the south-west. Its source region is the sub-tropical Atlantic Ocean, typically the Azores area, although occasionally it may come almost directly from the Caribbean. During its passage across the Atlantic, the air is cooled from below as it passes over a progressively cooler ocean. While it cools down, little of its moisture is lost. It therefore reaches south-west England or western Ireland almost saturated, giving dull, warm, overcast weather.

On the coast, sea fog is common in these tropical maritime southwesterlies. However, if the cloud base of the stratus or stratocumulus is several hundred metres, sea level sites may be saved from the fog, but on rising ground and hills there will be fog and drizzle. Bodmin Moor, Dartmoor, Dyfed, western Ireland and western Scotland can be shrouded in mild, damp conditions in winter and summer.

Further inland, in summer, the low stratus cloud may be burnt off by the Sun and it may become quite warm, although still humid. In the lee of hills or mountain ranges the clouds sometimes break up and there is a lot of sunshine. Favoured locations include Somerset, North Wales, Northumberland and the Moray Forth.

In a tropical maritime air mass the nights are mild and damp, especially in mid-winter. In December and January the overcast skies result in there being little variation in temperature between day and night. However, if there are light winds and clear skies, fog may form inland overnight.

d) Polar maritime

Polar maritime air is the most common type of air mass affecting the British Isles. The air has its source in the Canadian Arctic or the Greenland area. It reaches the British Isles from the west or north-west after having swung around the western side of a depression. As the cold air travels over the relatively warm sea it is warmed from below and becomes unstable. Unstable air streams tend to produce convection, and so cumulus clouds, cumulonimbus clouds and showers are likely in polar maritime air. Other characteristics of the air are that it is cool (especially in summer), fairly moist and associated with good visibility.

In winter most of the convection is initiated over the Atlantic and showers hit the coasts, spreading inland if the winds are strong. The Scottish and Welsh mountains often shelter the eastern side of Britain, although with a north-westerly wind some showers reach Birmingham and perhaps London. In spring and summer convection clouds tend to be set off inland by daytime heating. Such showers or short-lived thunderstorms can occur almost anywhere. At night the clouds disperse.

After a low has crossed eastwards over the British Isles, winds 'veer' (a clockwise change in wind direction) to a northerly point and true arctic air may reach Britain. This is sometimes referred to as arctic maritime air. It is similar to polar maritime air, but tends to be more unstable, colder and drier. Consequently showers of rain, snow, sleet or hail often occur on northern coasts and over high ground.

2 Low pressure systems

When two different air masses meet they form a front. For example, when a Pm and a Tm air mass converge the temperature differences between them may be over 10°C. The difference in density will allow the warmer air mass to rise over the cooler one. In any low pressure system (depression or cyclone) there are a number of forces operating simultaneously:

- the mixing of the two air masses
- the Coriolis force
- divergence of air aloft in the upper regions of the troposphere.

The result of these forces is to drag air inwards to the centre of the low pressure system.

In a depression (also referred to as a 'low') the fronts move with the wind, so they usually travel from the west to the east. The winds

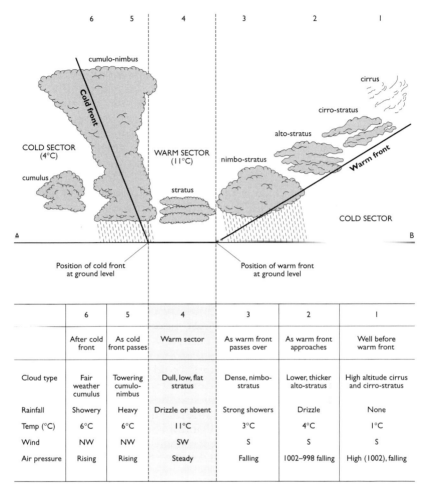

Figure 17 Cross section through a depression

	6	5	4	3	2	1
	After cold front	As cold front passes	Warm sector	As warm front passes over	As warm front approaches	Well before warm front
Cloud type	Fair weather cumulus	Towering cumulo-nimbus	Dull, low, flat stratus	Dense, nimbo-stratus	Lower, thicker alto-stratus	High altitude cirrus and cirro-stratus
Rainfall	Showery	Heavy	Drizzle or absent	Strong showers	Drizzle	None
Temp (°C)	6°C	6°C	11°C	3°C	4°C	1°C
Wind	NW	NW	SW	S	S	S
Air pressure	Rising	Rising	Steady	Falling	1002–998 falling	High (1002), falling

blow in an anticlockwise direction. At a front, the heavier cold air undercuts the less dense warm air, causing the warm air to rise over the wedge of cold air. As the air rises there is cooling and condensation, thus leading to the formation of clouds. If the cloud becomes sufficiently thick, rain will form. Consequently, fronts tend to be associated with cloud and rain (Figure 17). In winter there can be sleet or snow if the temperature near the ground is close to freezing.

In general, the appearance of a warm front is heralded by high cirrus clouds. Gradually, the cloud thickens and the base of cloud lowers. Altostratus clouds may produce some drizzle while at the warm front nimbostratus clouds produce rain. A number of changes occur at the warm front. Winds reach a peak and are gusty; temperatures suddenly rise; and pressure which had been falling remains more constant. Ahead of the warm front is a belt of thickening cloud, gradually developing into moderate rain and cloud. The belt of rain extends 160–320km ahead of the front. Behind the front the rain usually becomes lighter, or ceases, but the weather remains cloudy. As a warm front passes, the air changes from being fairly cold and cloudy to being warm and overcast (typical of warm air from the tropics travelling over the sea). Also there is a clockwise change in wind direction, and the wind is said to 'veer'. The warm front does not necessarily bring higher temperatures. This is particularly so in summer, when the cloudy weather behind the front cuts off the heat from the Sun. In winter, however, the south-westerly winds that usually blow after a warm front bring milder conditions. The advance of a warm front is usually the way in which a cold winter spell is broken down.

The cold front is marked by a decrease in temperature; cumu-lonimbus clouds and heavy rain; increased wind speeds and gustiness; and a gradual increase in pressure. After the cold front has passed, the clouds begin to break up and sunny periods are more frequent, although there may be isolated scattered showers associated with unsta-ble Pm air. A cold front moves so that the cold air is advancing to replace the warm air. This means that as a cold front passes, the weather changes from being mild and overcast to being cold and bright, poss-ibly with showers. The cold front usually brings a narrower belt of cloud and rain. It is called a cold front because the moist south-westerly winds ahead of it are replaced by cooler, drier north-westerly winds. When a cold front moves through an area it usually brings brighter, clearer weather behind it, but this brighter weather is sometimes mixed with showers.

The characteristics of an *occlusion* are similar to those of a cold front in that the rain belt is narrow, and the winds generally veer to the north-west behind it. There is usually a clearance to the west after the front has moved through. An occluded front occurs as the warm and cold fronts meet. Consequently, ahead of an occlusion the weather is similar to that ahead of a warm front, whereas behind the occlusion it is similar to that behind a cold front.

No two low pressure systems are the same. The weather that is found in any depression depends on the air masses involved. The greater the temperature difference between the air masses involved, the more severe the weather. Depressions are divided into **ana** and **kata** depressions depending upon the vigour of the uplift of warm air. The standard model of a depression was devel-oped by Bjerknes in 1937. Where air masses of differing composition meet, an *anafront* is formed and this produces cloud systems of great height. In contrast, *katafronts* occur when the air masses are fairly similar in composition. The formation of the precipitation is still largely explained in terms of the Bergeron-Findeisen process (1935) and the coalescence theory often working together simultaneously. In the Bergeron-Findeisen process, ice nuclei in clouds accumulate more ice by the condensation and freezing of the water vapour from surrounding supercooled droplets. (water vapour below 0°C, but still liquid). These then fall and melt to form raindrops. In the coalescence process larger water droplets falling fast under gravity collide with small ones and capture them.

3 High pressure systems

High pressure systems or *anticyclones* act very differently compared with low pressure systems. An anticyclone is a large mass of subsiding air, which causes high pressure at the surface (Figure 19). On a global scale, air moving to the poles from tropical areas sinks to form the

Case study – The Great Gale of 1987 and the great storm of 1990

The Great Gale of 1987 was one of the most severe storms ever to hit the south of the British Isles. It started as a low pressure system off the coast of Spain but was dragged north-eastwards to the British Isles by high level jet streams (Figure 18). Some of its energy was supplied by warm waters from the Bay of Biscay.

This was the strongest gale to affect much of the British Isles for 250 years, bringing with it gusts of up to 104 mph. The worst gusts were during the night, had they been during the daytime the effects would have been much greater (Table 6). The total costs of insurance claims were over £860 million.

Although the 1987 storm had a recurrence interval of 250 years (meaning that, on average, we would not expect a similar storm for another 250 years), there was another severe storm in January 1990. This storm was called a 'bomb', a term used to describe low pressure systems which intensify by 24Mb in 24 hours. Like the 1987 gale it was closely related to the location of jet stream activity.

The 1990 storm developed over the Atlantic and was easier to detect and predict than the 1987 storm. It affected a much larger part of Britain, with gusts of up to 74 mph. Like the 1987 gale it was a low frequency, high magnitude event, but the effects were different.

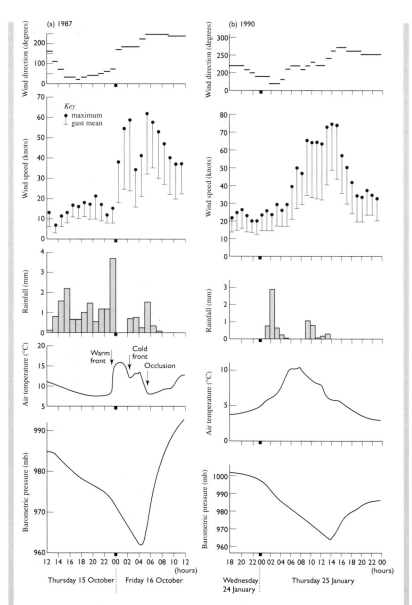

Figure 18 Weather data for a) the 1987 storm and b) the 1990 storm

	16 October 1987	25 January 1990
Meteorological parameters		
Fall in pressure	30Mb in 30 hours	39Mb in 24 hours
Maximum gust	100 knots	93 knots
Return period of an event of this magnitude	Greater than 200 years	150 years
Area of Britain affected		
Onshore	South-east England	Most of southern Britain, especially the south-west
Offshore	English Channel and North Sea	Most sea areas around Britain
Timing	Middle of night	Daytime
Season	Autumn	Winter
Effects		
Human fatalities	19	47
Trees blown down	15,000,000	5,000,000
Insurance claims	£1,117,000,000	Considerably more
Other effects	Worst power failure in south-east England since 1945 3000 miles of telephone lines brought down	Several hundred thousand homes without electricity
Summary	'The worst, most widespread night of disaster in the south-east of England since 1945' (Home Secretary, Douglas Hurd) The timing meant that the weekend was available for restoration of services, thereby limiting the disruption of commerce, industry and schooling.	Greater damage and death toll than 1987 on account of: • wider area affected • daytime occurrence • significance of successive, sustained gusts of high strength • progressive weakening of buildings and structures • restoration hampered by bad weather

Table 6 A comparison of the storms of 1987 and 1990

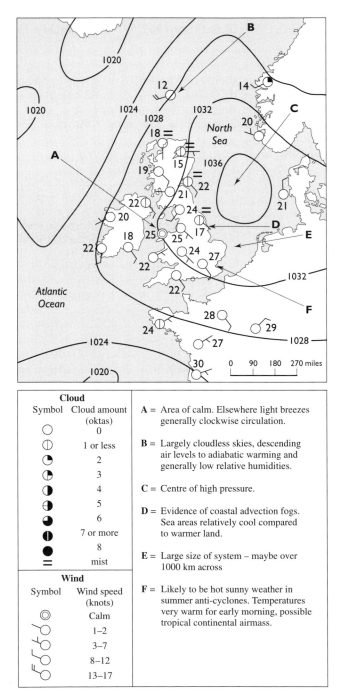

Figure 19 A synoptic chart of a summer anticyclone centred over the North Sea

sub-tropical high pressure belt, between 25° and 35° N and S. The other global high pressure areas are located in the polar regions.

Whereas low pressure systems produce wet, windy weather, high pressure systems produce hot, sunny, dry calm days in summer and cold, sharp, crisp days in winter. Nights are cold in winter as the lack of cloud cover allows heat to escape. Frost and fog are common in winter and autumn. Winds in a high pressure system blow out from the centre of high pressure in a clockwise direction (compared with a low pressure system where they blow into the centre of low pressure in an anticlockwise direction). The winds are light hence the isobars on a high pressure weather chart are circular and spaced far apart.

Most people look forward to high pressure systems because they bring warm, dry stable conditions, however, they can also cause problems. There are a number of hazards associated with high pressure systems, notably fog, low level ozone, and increasingly skin cancer and cataracts.

a) Blocking anticyclones

Within the upper westerly airflow (the *circumpolar vortex*) there are variations from a smooth westerly flow to a meandering pattern with large meanders and cut-offs in the air flow. These cut-offs produce cyclonic (low pressure) conditions in normally stable high pressure regions, and anticyclonic high pressure weather systems in normally low pressure regions. They are slow moving or stationary, and may persist for up to 5–6 weeks, hence the term blocking anticyclones.

Blocking anticyclones bring abnormal weather conditions to many places. Extremes of warmth, cold, drought, or flood may occur if the blocking system lasts long enough to dry out or saturate the soil. In 1976 in the British Isles temperatures of over 30°C were recorded, and in Finland in July 1972 it reached 33°C.

b) Atmospheric stability and instability

Air stability and instability refers to the buoyancy characteristics of air. High pressure is associated with stable conditions (stability) or **descending air** whereas low pressure systems are characterised by unstable conditions (instability) or **rising air**.

Instability results when a parcel of air is warmer and therefore less dense than the air above, causing it to rise and expand. This instability is the main cause of precipitation.

Stable air conditions (stability) exist when the ELR is less than both the DALR and the SALR. If a parcel of air is displaced upwards, it immediately gets cooler and denser and sinks. Uplift cannot be sustained. Conditions in an anticyclone are usually stable. The only time when stable air can rise is when it is forced over high ground.

c) Spatial variations in high pressure

A large proportion of the world's surface experiences prolonged high pressure. Semi-arid areas are commonly defined as having precipitation of less than 500mm per annum, while arid areas have less than 250mm and hyper arid areas less than 125mm per annum. Arid conditions are caused by a number of factors. The main cause is the global atmospheric circulation. Dry, descending air associated with the **sub-tropical high pressure** belt is the main cause of aridity around $20°–30°$ N. In addition, distance from sea, (**continentality**) limits the amount of water carried into these areas by winds. In other areas, such as the Atacama and Namib deserts, cold offshore currents limit the amount of condensation into the overlying air. Other arid regions are the result of intense rain shadow effects, as air passes over mountains. This is certainly true of the Patagonian desert. A final cause, or range of causes, are human activities. Many of these have given rise to the spread of desert conditions into areas previously fit for agriculture. This is known as desertification, and is an increasing problem.

Summary

- The map of global air pressure may be misleading. 'Average' pressure hides the fact that some areas have alternating periods of highs and lows.
- Some high pressure systems may remain for weeks, these are known as blocking anticyclones.
- Air masses are large bodies of air with relatively uniform temperature and moisture characteristics. They are crucial for the development of fronts, but are also associated with particular weather conditions if only one air mass dominates.
- Frontal weather occurs when two different types of air mass meet. The greater the contrast in air mass characteristics the more powerful the low pressure system.
- Low pressure systems typically produce wet, windy weather, sometimes gales may be produced as in the cases of the 1987 and 1990 storms.
- High pressure systems generally produce clear calm conditions, although these can be treacherous in winter, and may be associated with poor air quality too.

Questions

1. **a)** What is an air mass?
    b) Describe the main characteristics of the following air masses:
    - Polar continental (Pc)
    - Tropical maritime (Tm)

- Polar maritime (Pm)
- Tropical continental (Tc).

c) Where do polar maritime and tropical continental air masses come from? When are they most likely to affect the British Isles? What sort of weather do they bring?

2. Describe and explain the weather associated with a typical low pressure system.

Advice

1. a) An air mass is a large body of air, often thousands of kilometres wide, with fairly constant temperature and humidity characteristics
b) Pc – cold and dry
Tm – warm and wet
Pm – cold and wet
Tc – hot and dry
c) • North Atlantic and Sahara respectively
• winter and summer respectively
• cold and wet conditions; hot, dry conditions respectively.

2. This question is in two sections – describe and explain. Each is equally important. It's worth stressing that there is no such thing as a typical depression, although there is a model of one (rather like models of typical land use in a city – no city ever fits perfectly).

The description requires five sections
• ahead of the warm front
• at the warm front
• in the warm sector
• at the cold front
• after the cold front

Answers should include temperature changes, cloud cover and height, rainfall intensity and amount, wind speed and pressure.

An explanation requires a discussion of the mixing of cold air and warm air, the Coriolis effect, and reference to the role of jet streams and Rossby waves, (see Chapter 1, page 8). The use of case studies such as the gales of 1987 and 1990 provide excellent case studies to gain extra marks.

5 Microclimates

So far we have looked at global and regional climates. In this chapter we look at microclimates. These are small-scale, local climates found within a larger climatic region. For example, within cool, temperate climates there are distinct urban, forest and coastal climates caused by variations in heating, albedo and proximity to the sea. However, microclimates are only noticeable during high pressure conditions, since in a low pressure system winds mix the air and remove any differences that may have existed.

Microclimates have a vertical scale of tens of metres, a horizontal scale of kilometers, and a time scale of usually less than a day. They are characterised by a large difference in daytime heating and night-time cooling, and velocity varies from full velocity aloft to lower values and more variable speeds at ground level.

1 Urban microclimate

Urban microclimates occur as a result of extra sources of heat released from industry, commercial and residential buildings as well as from vehicles. In addition, concrete, glass, bricks and tarmac, all act very differently from soil and vegetation. Some of these, notably dark bricks, absorb large quantities of heat which they release slowly at night. Every new building modifies the existing climate. The release of pollutants also helps trap radiation in urban areas. Consequently, urban microclimates can be very different from rural ones (Table 7).

Three main changes occur relating to:

- atmospheric composition
- energy
- surface roughness and composition.

a) Atmospheric composition

Urban pollution reduces the amount of incoming solar radiation and increases the number of condensation nuclei (*hygroscopic nuclei*). The materials added to the atmosphere include aerosols and gases. The

Atmospheric composition	
Carbon dioxide (CO_2)	×2
Sulphur dioxide (SO_2)	×200
Nitrogen oxide (NOx)	×10
Carbon monoxide (CO)	×200
Total hydrocarbons	×20
Particulate matters	×3 to ×7
Radiation	
Solar	−15% to −20%
Ultra violet (winter)	−30%
Sunshine duration	−5% to −15%
Temperature	
Winter minimum (average)	+1−+2°C
Wind speed	
Annual mean	−20% to −30%
Number of calms	+5% to +20%
Fog	
Winter	+100%
Summer	+30%
Cloud	+5% to 10%
Precipitation	
Total	+5 to 10%
Days with <5mm	+10%

Table 7 Average urban climatic conditions compared with surrounding rural areas

main aerosols are carbon, lead and aluminium compounds. Most come from vehicle exhausts, industry and the burning of fossil fuels. The amount of small nuclei (0.01–0.1 micrometers diameter) is on average 9500cm³ in rural areas, but up to 4 million cm³ in urban areas. Large nuclei (0.5–10 micrometers diameter) have been shown to vary from 1–2cm³ in rural areas to 25–30cm³ in urban areas.

The main concentrations occur during periods of low wind speed and high relative humidities, when air is moving away from sources of pollution (e.g. dense concentrations of factories or fuel burning areas). There is a marked daily and seasonal pattern; peak concentrations occur at about 8 am in early winter. Pollution trapped beneath a temperature inversion raises the aerosol count. Smog (pollution and fog) can reduce insolation by over 50 per cent. The effect of smog is highlighted by conditions before and after London's Clean Air Act of 1956. Before the Clean Air Act there was a distinct difference between rural and urban areas. The use of smokeless fuels after the Clean Air Act cut London's emission of smoke from 141,000 tonnes in 1952 to 89,000 tonnes by 1960.

Gases such as sulphur dioxide (SO_2) are produced by industrial and domestic coal burning, and vehicles produce carbon monoxide (CO), hydrocarbons, oxides of nitrogen (NOx) and ozone (O_3). Before the Clean Air Act domestic fires in London produced up to 90 per cent of London's smoke but only 30 per cent of its SO_2. Most SO_2 came from electricity generating power stations (over 40 per cent) and factories (29 per cent). In Los Angeles, levels of pollutants are very high. Over 7 million people use 4 million cars, using 30 million litres of petrol daily, and producing 12,000 tonnes of pollutants. Up to 2.5 million litres of aviation fuel are used close to the city, and 13,500 lorries use 0.5 million litres of diesel each day. Despite guidelines, 7 per cent of the petrol used is emitted unburned, a further 3.5 per cent as photochemical smog and 33–40 per cent as CO. Los Angeles' smogs, unlike London's, occur during the day, usually in summer and autumn. Clear skies, light winds and temperature inversions combine with high levels of solar radiation to produce ozone.

b) Energy

Heat production due to human activity (factories, heating, transport, etc.) may exceed net radiation during winter in some cities. Although the radiation received in urban areas is not significantly different from rural areas (unless there is large-scale pollution) heat storage in urban areas is greater, leading to greater night-time temperatures. Evaporation rates in some parts of the city are very low, hence heat is transferred to the atmosphere rather than converting water from one state to another. This raises the temperature of urban areas. In contrast, well watered suburban lawns show a lower diurnal range in temperature as more of the heat is used up evaporating the water during the day.

The higher temperatures that characterise urban areas are the result of:

- changes in the energy budget due to the urban atmosphere
- changes in the energy balance due to the albedo and heat storage capacity of urban areas
- artificial sources of heat e.g. factories, vehicles and homes
- the reduced need of energy for evaporation.

Urban heat budgets differ from rural ones. During the day the major source of heat is solar energy; in urban areas brick, concrete and stone have high heat capacities. 1 kilometre of an urban area contains a greater surface area than 1 kilometre of countryside, therefore a greater area is heated in urban areas. For example, studies in Detroit, USA, showed that the urban atmosphere received 9 per cent less solar energy than nearby rural areas (25 per cent less during calm conditions). Although pollution reduces the incoming short-wave radiation, this is offset by the lower albedo and greater surface area in urban areas. The multiple surfaces of urban areas, and multiple reflection of short-wave radiation, allows for greater absorption of the smaller amount of energy.

Large urban areas produce as much heat in winter as they receive from solar radiation. The Boston–Washington megalopolis, covering an area of about 30,000km², with over 56 million people, produces heat equivalent to 50 per cent of winter radiation. In extreme climates, such as the Arctic, this proportion is even greater.

Urban areas are also characterised by a *heat island* effect. This is an island-shaped area of higher temperatures located over the city centre, declining outwards to the suburbs and rural-urban fringe. The heat island is clearest at dawn during high pressure conditions, where the cooling of rural areas is far greater than urban areas. The urban heat island may be 5–6°C warmer than the surrounding rural area. Between 1931 and 1960 the centre of London had a mean annual temperature of 11°C compared with 10.3°C in the suburbs, and 9.6°C in the surrounding rural areas. It is thought that domestic fuel contributes between one-third and one-half of the additional heat in the urban area.

Heat islands are most noticeable when wind speed is low, below 5–6cm/s. In the absence of regional winds, the heat island may develop its own circulation with winds blowing inwards to the warmer, lower pressure zone over the city centre. London's heat island is most noticeable during summer when pollution and heating are at a minimum. This suggests that the storage of heat by buildings, and subsequent release, is one of the most important factors causing the extra heating. As a result of the heat island, urban areas have fewer frosts. In London, Kew has 70 more frost free days than nearby (rural) Wisley.

The typical heat profile of an urban heat island will show the maximum temperature at the city centre, a plateau across the suburbs and a temperature cliff between the suburban and rural area (Figure 20).

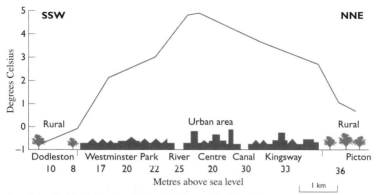

Source: Briggs, D. et al., *Fundamentals of the Physical Environment*, Routledge, 1997

Figure 20 Temperature cross-section of an urban heat island

Small-scale variations within the urban heat island occur due to the distribution of industries, open space, rivers, canals, etc. Studies have shown that the contrast in temperature between rural and urban areas depends on the size of the urban area. For example, in North America:

- urban areas with 1000 people were, at most, 2.5°C warmer than rural areas
- urban areas with 100,000 people were, at most, 8°C warmer than rural areas
- urban areas with 1,000,000 people were, at most, 12°C warmer than rural areas.

The nature of urban climates is changing. With the decline in coal as a source of energy there is less SO_2 pollution, resulting in less hygroscopic nuclei and less fog. However, an increase in cloud cover has occurred for a number of reasons:

- greater heating of the air (forcing the air to condense)
- increase in pollutants
- frictional and turbulent air flow
- changes in moisture content.

c) Airflow

In general, wind speeds are lower in urban areas than rural areas due to sheltering by buildings. Wind speeds in the city centre are usually 50 per cent lower than the suburbs; however, buildings have a major impact on airflow causing turbulence and channelling.

The effect on airflow varies seasonally and daily. During the day, wind speeds in the city are lower than rural areas, but by night the differences are less. For example, surveys at London airport (a suburban site) and central London recorded average wind speeds of 2.9m/s and 2.1m/s respectively.

d) Moisture

The lack of large water bodies and the speedy removal of rainfall through gutters, drains and storm channels, reduces the amount of water present, and therefore the amount of evaporation in urban areas. The lack of vegetation cover reduces evapotranspiration which also reduces the absolute humidity over urban areas.

Conversely, thunderstorms are more frequent over urban areas. Between 1951 and 1960, thunderstorms in the south-east of the British Isles were concentrated in south, west and central London. Moreover, over this period rainfall in London's thunderstorms was up to 25cm more than in the rural areas (owing to higher temperatures and greater convectional heating). Parts of London received between

25 per cent and 50 per cent more rain compared with other areas in the South East.

e) Tropical urban climates

Most of the world's very large cities ('*supercities*') are in less economically developed countries (LEDCs). Most LEDC urban land use differs from more economically developed countries (MEDCs) in that it is normally high density, single storey buildings, fewer open spaces, and poor drainage. Tropical heat islands are less pronounced than temperate heat islands, largely as a result of less human-generated heat sources. However, in newly industrialising countries (NICs), with their emphasis on industrialisation, sources of heat are similar to MEDCs earlier in the twentieth century.

Urban climates in Mexico have received much attention. The heat island is best developed in the dry season (November–April), with its high pressure, low wind speeds, clear skies and temperature inversions. In coastal areas, such as Veracruz, the heat island effect can be so pronounced as to draw in air from the sea. This cools Veracruz making it cooler than the surrounding rural areas.

2 Microclimates over bare soil

There are important variations in the microclimate caused by soil factors. Humus rich (dark) soils absorb heat more than light soils, and moist soils warm up more slowly than dry soils due to the specific heat capacity of water. Soils that have a high proportion of air are poor conductors of heat, and hence sandy soils may be very hot by day at the surface but cool down rapidly with depth, and at night they become very cold. In contrast soils with some moisture can transfer heat downwards. If there is too much water in the soil (over 20 per cent) it may take much longer to heat up the soil.

Under clear conditions at night, long-wave radiation losses from the soil surface exceed the return of radiation from the atmosphere. In contrast when conditions are cloudy, water vapour absorbs long-wave radiation and returns much of it to the ground. On windy days air is mixed and so temperature variations are minimal, but under high pressure conditions night-time temperatures are very low and day-time temperatures very warm. The factors promoting the maximum difference day and night-time temperatures are:

- clear skies
- dry air
- lack of wind
- sandy soils
- snow covered ground.

3 Microclimates over vegetated surfaces

a) Forest climates

Trees and forests can have a marked effect on climate. Air movement is much less within a forest; temperatures are lower and forests are more humid than open land. Most of the incoming solar radiation is absorbed by the *canopy layer* (the top layer of trees), although some energy is reflected (the albedo varies between species). This means that only a small proportion of incoming radiation reaches the ground, hence forests do not heat up as much as open areas, during the day. At night, vegetation traps and returns much of the outgoing long-wave radiation, the presence of water vapour also helps absorb long-wave radiation at night.

The vertical structure of forests largely determines their microclimate. The structure is determined by species composition, size, coverage and layering. Other important characteristics include seasonality (evergreen versus deciduous), and the size and density of the leaves. There are marked seasonal differences; deciduous forests lose their leaves in winter resulting in much less absorption and interception. There are important differences between the type of forests:

- deciduous trees create larger seasonal differences than evergreens
- large-leaved trees such as sycamore absorb more energy than small-leaved trees, e.g. birch and oak
- oak trees have a higher density of leaves than birch trees hence more light reaches the ground in a birch wood.

In tropical rain forests, the average height of the canopy species is 45–55m, with some species over 60m. In contrast, in temperate forests average height is only 30m. Tropical forests have a greater variety of species, 100 per km² compared with under 25 (and sometimes only one) in temperate forests. Temperate forests often have just two or three layers (field, shrub and canopy) compared with five in tropical forests. If a forest is layered, there will be additional interception at each layer, and the outgoing long-wave radiation will be absorbed at each layer. In a tropical rain forest, up to 99.9 per cent of the energy available is absorbed by the trees, less than 0.1 per cent reaches ground level. Vapour pressure is higher in a forest than in open land, because of the presence of large amounts of moisture from the leaves and the low rates of evaporation due to cooler tempatures and low wind speeds.

Forest canopies alter the pattern of insolation and reradiation. Coniferous forests have albedos between 8 and 14 per cent and deciduous woodlands 12–18 per cent. For a dense forest of red beeches, 80 per cent of incoming radiation is trapped by the canopy and less than 5 per cent reaches the forest floor. Greater interception occurs in sunny conditions. Interception also varies with age. Studies of Scots

Pine have shown the following differences in the amount of solar radiation reaching the forest floor:

- at 1–3 years 50 per cent reached the forest floor
- at 20 years 7 per cent reached the forest floor
- at 130 years 35 per cent reached the forest floor.

Air flow in forests is much lower than in open spaces. Studies of European forests show that:

- at 30m into the forest, wind speed was 60–80 per cent of that of open ground
- at 60m it was 50 per cent
- at 120m it was 7 per cent.

An understanding of the way in which forests modify air flow has been used to design wind breaks. The denser the wind break, the greater the shelter it provides, although some turbulence may be created down wind. The reduction of wind speed increases the risk of frost. In addition, forests can filter dust and fog droplets from the air. Under certain conditions, so much fog is removed from the air that there is more precipitation inside a forest than outside it.

Humidity conditions inside the forest are very different from outside the forest. Evaporation in the forest is low due to reduced light intensity, low wind speeds, lower maximum temperatures and high air humidity. The humidity of the forest depends on the amount of evapotranspiration and increases with the amount of vegetation. Forest humidity is generally about 10 per cent higher than in the open.

Forests have an important influence on temperature. This is due to:

- shelter from the Sun
- heat loss by evapotranspiration
- blanketing at night
- reduced air flow.

Consequently, daytime maximum temperatures are lower and night-time minimum temperatures are higher than surrounding areas. Mean annual temperatures are about 0.6°C lower than the surrounding area, and temperature differences are generally greater in summer (2.2°C) compared to only 0.1°C in winter.

b) Crops

The microclimate of short green crops shows a number of characteristics:

- maximum temperature occurs just below the top of the vegetation in early afternoon; this is the point where maximum absorption occurs. Temperatures are lower near the soil where heat is transferred into the soil

- wind speed is at a minimum in the upper crop canopy, where the vegetation is most dense
- the maximum evaporation rate occurs where the crop canopy is densest, hence humidity is greatest at this point
- the maximum amount of CO_2 is found at night in the crop canopy (due to respiration) while CO_2 is absorbed in photosynthesis during the day.

c) Hedges

A 2m high hedge can result in reduced wind speeds for as much as 56m beyond it. The maximum decrease (up to 40 per cent of the original speed) occurs approximately 8m beyond the hedge. This dramatic effect makes hedges extremely important in providing shelter for livestock. The planting of long, thin blocks of trees as shelter belts has a similar effect over even longer distances. A significant reduction in soil erosion occurs as a result of the protection offered by hedges.

Hedges have a number of other microclimatic effects. Soil moisture content and the daytime air and soil temperatures can be increased by as much as 16–20 per cent in the lee of a hedge, with the effect reaching as far as 10 times the height of the hedge. Evaporation can be significantly decreased at a distance of 15 times the height of the hedge.

4 Local winds

a) Valley breezes

Katabatic winds are winds which blow down the slope or valley at speeds of about 1m per second. These typically occur at night as cold air drains down a mountain. During the day the heating of the valley floor causes air to rise, and the consequent upslope breezes are known as *Anabatic winds* (Figure 21).

b) Land–sea breezes

Sea breezes are formed by the differences in the specific heat capacity of land and sea. During the day, under calm conditions, the land heats up quickly as it absorbs short-wave radiation, and the ground in turn heats the air above it. At night, the land cools rapidly since there is little stored heat (the land is a poor conductor of heat and does not transport heat downwards efficiently).

In contrast, incoming radiation can reach a depth of about 30m in the sea; water has a high specific heat capacity, thus it needs much energy to raise its temperature. In addition, tides and currents mix the heat over a large area, thus the sea gains heat slowly by day and releases it slowly by night. During the day, a weak high pressure system develops over the cool sea while a low pressure system develops

Night-time mountain breeze

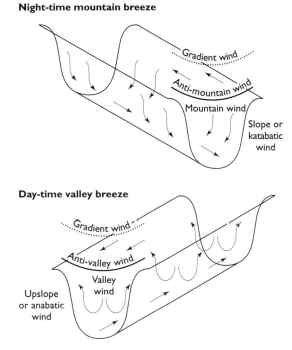

Day-time valley breeze

Source: Briggs, D. et al., *Fundamentals of the Physical Environment*, Routledge, 1997

Figure 21 Mountain and valley breezes

over the heated land. Winds blow from the high pressure over the sea to the low pressure over the land. These winds may reach up to 7m/s. As the sea breeze moves inland it brings cool, humid air to hot, dry areas inland. At night, the wind flows from the land to the sea, but the speed is usually lower, typically 2–4m/s, due to less energy being available.

Summary

- Microclimates are small-scale climates, and are only noticeable during calm high pressure conditions.
- Urban microclimates often display a heat island, poor air quality, changes in air flow, and changes in the amount and type of precipitation.
- Bare soils also exhibit microclimates. Much depends on the moisture content of the soil.
- Microclimates associated with vegetation vary with the type of

vegetation. Forest microclimates depend on the type and size of the forest; crops and hedges also modify the climate although to a smaller extent.

• In other areas, microclimates are associated with particular wind systems. In mountain areas, winds blow up the slopes by day, but by night cold air drains down the slope. In coastal areas, land–sea breezes develop where winds blow from the cooler sea to the warmer land by day, but by night the flow is reversed.

Questions

1. Study Figure 22 which shows the effect of a sea breeze on the Midlands.
   a) Approximately what time did the sea breeze arrive?

(b)

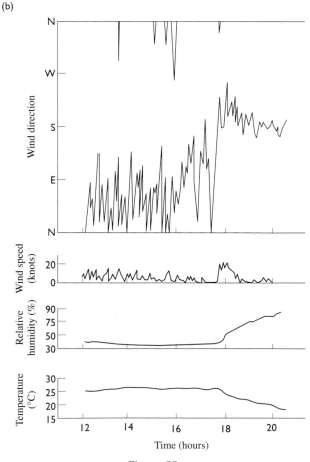

Figure 22

b) What was the main direction of the wind:
- before the sea breeze
- after the arrival of the sea breeze?

c) Suggest reasons why:
- the relative humidity increased after the arrival of the sea breeze
- why the temperature decreased.

d) Why are microclimates, such as sea breezes, best observed during high pressure (anticyclonic) weather conditions?

2. Using examples, describe and explain the conditions associated with two named microclimates.

Advice

A/S

1. **a)** 1800 hrs
  b) E-N; S
  c) Because air temperature dropped, the potential amount of water it could hold dropped thus the amount of moisture it was holding compared to the total it could hold increased; (b) the temperature dropped because it was influenced by the sea air – the air which was cooler due to the sea's higher specific heat capacity.
  d) During high pressure systems conditions are calm (low wind speed) and cloud cover is limited, so temperature differences between urban areas, or land and sea for example, are at a maximum, especially at night, and there is little wind to mix the air. In contrast, during low pressure conditions air is mixed by turbulent air flows hence the differences in temperature of air above land and sea are minimised.

A2

2. This is in two parts – describe and explain. A description should include temperature (maximum, minimum and range), precipitation (type and frequency) cloud cover and wind speed. Examples must be used. These should be real examples such as those in this chapter. The explanation requires an analysis of a number of points depending on the examples chosen. These could include:

- sources of heating (in an urban microclimate)
- albedo
- specific heat capacity
- sources of moisture (or lack of)
- effect of shading
- aspect
- funnelling of winds.

In a 20 mark question, a typical mark scheme might look as follows:

Level I   0–3 marks
Largely irrelevant; rambling account; a few points; no real focus. Candidates make a few generalisations about *one* microclimate, with very little supporting evidence. Most of the material is descriptive and there is no clear attempt to explain the processes involved.

Level II   4–7 marks
Descriptive account; some generalisations and order; very limited detail. Candiadates recognise that *two or more* microclimates are needed, and describe the main characteristics of two systems.

Level III   8–11 marks
Good description and use of examples; limited explanation. Candidates have clearly ordered their material – accurately describing variations in temperature, wind systems, precipitation where relevant. The explanation may be partial rather than detailed – e.g. extra heating in urban areas causes temperatures to be higher (rather than the extra heating from x and y causes *minimum* temperatures at night to be higher).

Level IV   12–15 marks
Good description and explanation; lacks the flair and punch of the most able candidates; detailed and thorough. Contrasts at least two microclimates and uses data to support evidence. Explanations are detailed and thorough – linking cause and effect.

Level V   16–20 marks
Very good – description, explanation and very good use of examples; compares places i.e. evaluates case studies. Candidate uses own data, e.g. school fieldword or data from articles. May attempt an overall evaluation, e.g. urban microclimates vary depending on population size, level of development volume or traffic, etc.

6 Climatic hazards

Natural hazards are a reminder to us of the power of nature and the vulnerability of human systems. It would appear that natural hazards are becoming worse – more frequent, more violent and having a greater impact – but this could be due to there being more people, more property at risk and better reporting. In this chapter we look at some of the issues related to natural hazards, and these are illustrated by a number of case studies.

1 Natural hazards

A hazard is a perceived natural event which threatens both life and property. A basic distinction can therefore be made between extreme events in nature, which are not environmental hazards (because people and/or property are not at risk) and environmental hazards in which people and/or property are at risk. Environmental hazards are caused by people's use of dangerous environments. A large number of environmental hazards are caused by human behaviour, namely the failure to recognise the potential hazard and act accordingly. Hence the term 'natural hazard' is not a precise description, as they are not just the result of 'natural' events.

Environmental hazards have a number of common characteristics:

*   the origin of the hazard is clear and produces distinct effects, such as flooding causing death by drowning
*   the warning time is short (although drought is an exception)
*   most losses to life and property occur shortly after the environmental hazard; these are often related to secondary hazards such as fire and contaminated water
*   in some areas, especially LEDCs, the risk of exposure is largely involuntary, normally due to people forced to live in hazardous areas; by contrast in most MEDCs people occupy hazard areas as much through choice as through ignorance or necessity
*   the disaster occurs with a scale and intensity that requires an emergency response.

It is possible to characterise hazards and disasters in a number of ways.

1. **Magnitude** The size of the event e.g. a force 10 gale on the Beaufort Scale, the maximum height or discharge of a flood, or the size of an earthquake on the Richter Scale.
2. **Frequency** How often an event of a certain size occurs, e.g. a flood of 1 metre height may occur, on average, every year. In the same stream a flood of 2 metres in height might occur only every 10 years. The frequency is sometimes called the recurrence interval.
3. **Duration** The length of time that environmental hazard exists. This varies from a matter of hours, such as urban smog, to droughts lasting decades.
4. **Areal extent** The size of the area covered by the hazard. This can range from very small scale, such as an avalanche chute, to continental (drought).
5. **Spatial concentration** This is the distribution of hazards over space. For example, they may be concentrated in certain areas, such as plate boundaries, coastal locations, valleys, etc.
6. **Speed of onset** This is rather like the 'time-lag' in a flood hydrograph. It is the time difference between the start of the event and the peak of the event. It varies from rapid events, such as the Kobe earthquake, to slow time scale events such as drought in the Sahel of Africa.

Country	Date	Deaths	People affected	Money pledged ($ million)	Economic cost ($ billion)
Mozambique	Mar 2000	400	2 million	107	n/a
Venezuela	Dec 1999	30,000	600,000	27.8	15.0
India (Orissa)	Nov 1999	10,000+	12 million	20.8	2.5
China	Aug 1998	3600	200 million	131.7	30.0
Bangladesh	Sep 1998	4750	23 million	234.1	5.0

(*Source: The Economist,* 11 March 2000)

Table 8 The impact of floods 1998–2000

7. Regularity Some hazards are regular, e.g. cyclones; others are much more random, e.g. earthquakes and volcanic eruptions.

Disasters which have killed over 500,000 people this century include drought in India (1900), drought in the Soviet Union (1921) and floods in China (1928, 1931, 1939) (Table 8). Since the Second World War there have been other massive killers; the Bengal cyclone (1970), the earthquake at Tanshen, China (1976), and droughts in Mozambique (1981) and Ethiopia (1984).

Famine in Africa, 2000

In 2000, up to 16 million people were at risk of starvation in the Horn of Africa. This was largely the result of prolonged drought as well as years of political conflict. The United Nations appealed for over £100 million to help with the crisis.

Up to 8 million Ethiopians were at risk of starvation, and many more in Eritrea, Kenya, Somalia, Uganda and Djibouti (Figure 23). Amongst the worst hit areas was Ogaden in eastern Ethiopia where it had not rained for almost three years. There, dozens of children were dying daily, from malnutrition, measles and TB.

There had been a number of 'natural' events, which had contributed to the famine:

- nearly three years of drought
- sporadic intense storms washing away soil
- frost
- infestations of black beetles
- hail.

The human conflicts were equally important, and more so with respect to the provision of food aid.

Even when food aid was provided there was a problem in delivering it. Civil war in Sudan, fighting in Somalia and the continued conflict between Eritrea and Ethiopia made food distribution difficult. The port of Massawa in Eritrea was closed and the only way for food to be brought into Ethiopia was through Djibouti, but that was too small to handle the volume of aid that was required.

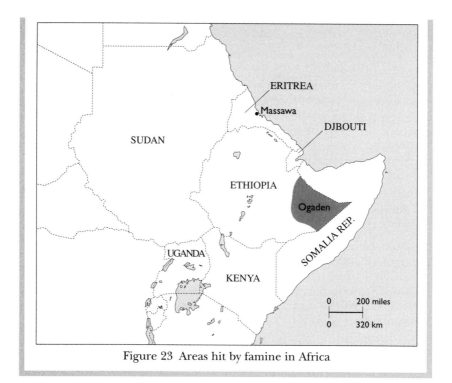

Figure 23 Areas hit by famine in Africa

About 90 per cent of the world's environmental hazards that cause more than 100 deaths (in a single event) are caused by four types of hazards:

- floods 40 per cent
- tropical cyclones 20 per cent
- earthquakes 15 per cent
- drought 15 per cent.

The loss of life and property experienced in a disaster is affected by a number of factors:

- size of hazard
- population density
- level of adjustment to hazard
- experience of previous hazards
- perception of hazard threat.

Environmental hazards claim more lives in LEDCs. Over 90 per cent of deaths related to environmental hazards occur in LEDCs. This is partly because poorer countries cannot afford disaster planning procedures or proper prevention methods. In contrast, 75 per cent of the economic damage occurs in MEDCs.

Storms in Europe – the start of increased storm frequency

The recent increase in storm frequency started between Christmas and the New Year, 1999–2000:

- in the British Isles five people were killed on the mainland and two in the Channel
- winds of over 180km/hour killed 33 people in France and uprooted 100,000 trees in Paris
- as a result of the storm, 3–4 million homes in France were without electricity (one-quarter of the country)
- nearly 750,000 French homes were without use of their telephone
- major archaeological treasures such as Versailles and Notre Dame cathedral were affected by the storm
- about £50 million worth of damage was done to Paris's historic monuments
- all trains in the south-west of France were halted
- in ports such as La Rochelle large yachts were destroyed and flung into the town
- three nuclear reactors in Blayais had to be closed down after water from the Gironde River flooded the buildings
- the storms also washed onshore oil from an oil spill
- in Austria avalanches killed nine people in Galtur
- eleven people were killed in Switzerland and another four in Italy due to avalanches
- EuroDisney was shut down and the hotels evacuated.

Much of the public attention was caught by the huge number of trees uprooted in France. Described as an 'ecological nightmare' 80 per cent of France's woodlands were affected. Some were historical woodlands, such as Fontainebleau near Paris, and others were 'working' forests such as those in Alsace and Lorraine. The livelihood of many thousands of people in those areas was also seriously affected.

In general, the impact of environmental hazards in MEDCs has a greater economic cost but causes fewer deaths. In contrast, in LEDCs, environmental hazards are causing an increase in the number of deaths as well as increasing economic costs. The increase in the number of deaths is due to a number of factors:

- population growth and the use of marginal (unsafe) land for dwellings
- a lack of land due to environmental deterioration
- economic growth creating new hazards such as chemical spills and radiation leaks
- technical innovations such as high rise flats and large dams.

Environmental hazards only occur when people and property are at risk. Although the cause of the hazard may be geophysical or biological this is only part of the explanation. It is because people live in hazardous areas that hazards occur. So why do they live in such places? The **behavioural** school of thought considers that environmental hazards are the result of natural events. People put themselves at risk by, for example, living in floodplains. In contrast, the **structuralist** school of thought stresses the constraints placed upon the (poor) people by the prevailing social and political system of the country. Hence, poor people live in unsafe areas (such as steep slopes or floodplains) because they are prevented from living in better areas. This school of thought provides a link between environmental hazards and the underdevelopment and economic dependency of many LEDCs.

At an individual level there are three important influences on a person's response to hazard:

* experience – the more experience of environmental hazards the greater the adjustment to the hazard
* material well-being – those who are better off have more choice
* personality – is the person a leader or a follower, a risk-taker or risk-minimiser?

Ultimately there are three choices:

1. do nothing and accept the hazard
2. adjust to the situation of living in a hazardous environment
3. leave the area.

Pingelap Atoll – the case of achromotopsia

In 1775 a typhoon swept the tiny Pingelap Atoll in the Pacific ocean. The island's crops and livestock were devastated and the survivors were almost entirely wiped out by famine. Yet 20 people pulled through the catastrophe and repopulated the island.

Four generations after the typhoon, a rare genetic disease called achromotopsia emerged. Sufferers were left with poor eyesight, total colour-blindness and painful sensitivity to sunlight. With such a tiny human population, the faulty gene was quickly passed around and today 5–10 per cent of the island's population of 3000 people suffer from the genetic disorder. All can trace their ancestry back to a single male survivor of the typhoon and medical researchers believe this man carried a recessive gene mutation for achromotopsia. It was only after his descendants inter-married that the genes were shuffled around and eventually the disease was revealed.

The fact that we know so much about the disease is due to the isolation of the islanders and a storm over 200 years ago.

The Venezuelan mudslides

The Venezuelan mudslides of 1999 were the worst disaster to hit the country for almost 200 years. The first two weeks of December saw an unusually high amount of rainfall in Venezuela. On 15 and 16 December the slopes of the 2000m Mount Avila began to pour avalanches of rocks and mud, burying large parts of the coastal *litoral central* (the Central Coast). About 350,000 people lived in this area.

Over 30,000 people were killed and about 200,000 people were made homeless. Most of the dead were buried in mudslides that were between 8 and 10m deep. The true number of casualties may never be known. The mudslides also destroyed roads, bridges, factories, buried crops in the fields, destroyed telecommunications, and also weakened Venezuela's tourist industry.

The disaster was not just related to heavy rainfall. The present government blamed corrupt politicians from previous governments and planners who had allowed shanty towns to grow up in the steep valleys surrounding the coast and the capital, Caracas.

Venezuela is not a poor country. It has a broad range of resources such as oil and gas, bauxite, iron ore and gold. It has a high life expectancy, a low infant mortality and a high literacy rate. The reason for the huge death toll was the fact that 85 per cent of the population live in urban areas, and the concentration of people in cities and shanty towns in the region affected by the mudslides increased the impact of the rain. Were people distributed more evenly around the country the impact would have been less tragic.

It is the adjustment to the hazard that we are interested in.

The level of adjustment will depend, in part, on the risks caused by the hazard. This includes:

- identification of the hazard
- estimation of the risk (probability) of the environmental hazard
- evaluation of the cost (loss) caused by the environmental hazard.

The adjustment to the hazard includes three main options:

1. **Modify the loss burden** – spread the financial burden, e.g. insurance, disaster relief
2. **Modify the hazard event** – tailor the building design, building location, land-use zoning, emergency procedures. Efforts that have been made to control extreme events include flood relief schemes, seawalls, avalanche shelters, etc.
3. **Modify human vulnerability to hazard** – install emergency procedures, forecasting and warning systems.

2 Hurricanes

The name hurricane should only be used for those tropical storms occurring in the Atlantic. In the Pacific they are known as typhoons, in the Indian Ocean as cyclones and in Australia as willy-willies. Each year they are given names beginning with 'A', 'B', etc., in order of occurrence and the names are alternately male and female.

Hurricanes are one of the most dangerous natural hazards to people and the environment. Every year immense damage is done by hurricanes and other similar tropical storms, such as Hurricane Mitch in 1998 and Hurricanes Eline and Gloria in 2000. However, hurricanes are essential features of the Earth's atmosphere, as they transfer heat and energy between the equator and the cooler regions towards the poles.

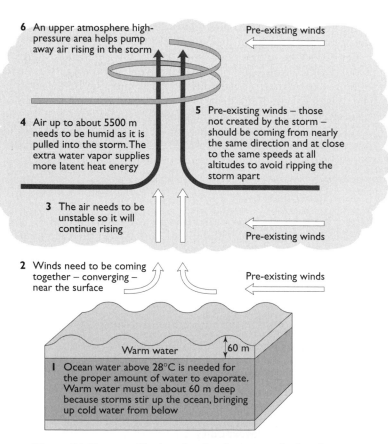

Figure 24 Factors affecting the development of a hurricane

A hurricane is a large rotating storm centred around an area of very low pressure with strong winds blowing at an average speed of over 114km per hour. The whole storm system may be 8–10km high and 480–640km wide (Figure 24). It moves forwards like an immense spinning top, at speeds up to 50km per hour.

There are several factors required to transform storms into hurricanes. The most important ones are:

- a source of very warm, moist air – derived from tropical oceans with surface temperatures greater than 26°C
- sufficient spin or twist from the rotating Earth – this is related to latitude.

As the warm sea heats the air above it, warm moist air rises quickly, creating a centre of low pressure at the surface. Winds rush in towards this low pressure and the inward spiralling winds whirl upwards, releasing heat and moisture before descending. The rotation of the Earth causes the rising column to twist. The rising air cools and produces towering cumulonimbus clouds. Further aloft at 8–10km altitude the cloud tops are carried outwards.

Large amounts of energy are transferred when warm water is evaporated from tropical seas. As this air rises up to 90 per cent of the stored energy (water vapour) is released by condensation, giving rise to the towering cumulus clouds and rain. The release of heat energy warms the air locally causing a further decrease in pressure. Consequently, air rises faster to fill this area of low pressure, and more warm, moist air is drawn off the sea feeding further energy to the system. Thus a self-sustaining heat engine is created.

Hurricanes form between 5° and 30° latitude and initially move westwards (owing to easterly winds) and slightly towards the poles. Many hurricanes eventually drift far enough to move into areas dominated by mid latitude westerlies. These winds tend to reverse the direction of the hurricane to an eastwards path. As the hurricane moves polewards it picks up speed and may reach 32–48km per hour. An average hurricane can travel about 480–640km a day, or about 4800km before it dies out.

In the northern hemisphere, hurricanes occur between July and October in the Atlantic, eastern Pacific and the western Pacific. South of the equator, they occur between November and March off east Australia and in the Indian Ocean.

The most common phenomena associated with hurricanes are strong winds. Other phenomena include:

- hurricane waves – large waves up to 15m high are caused by the strong winds and bring about extensive flooding.
- swells – an increase in ocean level.
- rain – the hurricane picks up about 2 billion tonnes of moisture per day and releases it as rain; in extreme cases up to 60cm of rain may fall in 36 hours.

These phenomena can cause major destruction, especially when the hurricane's path takes it over land. However, a path over land also causes the destruction of the hurricane itself. As it moves over land its energy source (water) is depleted and friction across the land surface distorts the air flow. This leads to the eye filling with cloud and the hurricane dies.

Other than a basic knowledge of general hurricane occurrence, there are no atmospheric conditions that can be measured to predict where and when a hurricane will develop. Therefore we can only forecast its path once formed. Satellites detect hurricanes in their early stages of development and can help to provide early warning of imminent hurricanes. Reinforced aircraft fitted with instruments fly through and over hurricanes, and weather radar can locate hurricanes within 200 miles of the radar station.

3 Flooding in Mozambique and the British Isles

a) Flooding in Mozambique, 2000–1

Mozambique is one of the world's poorest and most indebted countries. Until 1975 it was a Portuguese colony, and when the Portuguese

---

Flooding in South Africa, 2000

- More than 35 people were killed in flooding in South Africa during February 2000. Torrential rains swamped parts of southern Africa and major road links in the region were affected.

- The cause of the floods was intense rain and thunderstorms making most rivers rise to their highest levels in 50 years. Up to 445mm of rain fell in just four days in the Kruger National Park. The park's Skukuza camp was cut off from the outside world for the first time in its 101-year history.

- As ever, impoverished communities were hardest hit by the torrential rains. In Alexandria, north of Johannesburg, the Jukskei River burst its banks and at least 120 Alexandra shacks on the edge of the river were swept away. About 300 families were evacuated from low-lying areas in Kliptown and seven people were rescued from rising waterways.

- The cost to repair flood damaged bridges, roads and government buildings in Mpumalanga (South Africa) was estimated at £25–50 million.

- Business at the country's famed Kruger National Park was disrupted with the closure of several rest camps. Scores of tourists were evacuated from the park, disrupting one of South Africa's lucrative hard currency earners. Most of the animals were able to reach high ground.

left Civil War broke out, laying waste to much of the country. In 1992 Mozambique established a democratic government, and it had been developing and making good economic progress until the 2000–1 floods.

Between January and April 2000 flooding occurred throughout southern Africa. South Africa, for example, experienced major problems with flooding, although this was nothing compared with the problems that were later to affect Mozambique. Flood waters in South Africa contributed to the build up of water that was later to drain through Mozambique.

The floods that devastated Mozambique were caused by a long period of heavy rain. Over 1100mm of rain fell in February, indeed more than 75 per cent of Mozambique's annual rainfall fell in just three days (Figure 25). Heavy rain persisted for over four weeks and the rescue operation was hindered when even more torrential rain fell, following a spell of dry conditions. Two tropical cyclones, Eline and Gloria, brought driving rains and strong winds to the region, further increasing the problem.

The Mozambique floods continued for four months. This was partly due to the rain which kept falling, but it was also due to the fact that the rivers which flow through Mozambique, the Limpopo, the Save and the Zambezi, rise far away in South Africa and Zambia, and drain through Botswana and Zimbabwe. This meant that even after the rain had stopped in Mozambique it was receiving floodwaters from Zambia, Zimbabwe and Botswana. Initially these waters were ponded up behind dams, but eventually the dams could not contain the floods and had to allow the water through.

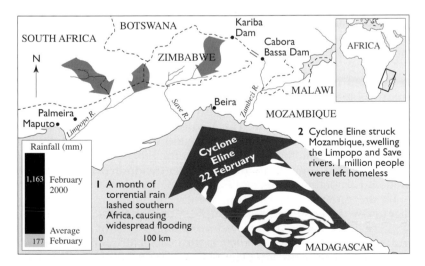

Figure 25 Heavy rainfall and flooding in Mozambique

The floods killed about 700 people, although the true number may never be known. Up to a million people were made homeless, and the floods ruined much of Mozambique's infrastructure (roads, bridges, shops and factories). In areas where the floodwaters fell by the end of March it was possible to plant some seeds for harvest. However, Mozambique continued to be affected by hurricanes well into April. The areas worst affected were the low-lying fertile valleys, which were the best areas for farming. In some areas the floodwaters remained high and it was not possible to plant any crops in 2000 so some farmers were dependent on food aid until 2001.

Even when the floodwaters dropped, the hazard was not over. Pools of stagnant water are ideal breeding grounds for mosquitoes, which cause malaria. People drinking contaminated water are at risk of cholera, diarrhoea and vomiting. Shortages of food affected peoples' nutritional state, and malnutrition became widespread. Other hazards included poisonous snakes and spiders washed up by the swollen rivers. Even land mines planted during the civil war were washed up by the floodwaters.

The main short-term needs were clean water, food, medicine and tents for the refugees. The long-term needs were the rebuilding of the nation's infrastructure, and the rebuilding of Mozambique's fragile economy. The aid effort, which has been criticised on account of the time it took to get started, has had some impact, albeit limited. However, because Mozambique is a sparsely populated country it was difficult to get the aid to the people.

In early April a further hurricane, Hurricane Hudah, killed at least 27 people and left about 100,000 people homeless in Madagascar. The worst affected area was the coastal town of Antalaha in the north-east of the country. Just before the hurricane hit the Mozambique coast it suddenly changed direction. It travelled along the coast rather than inland, and so Mozambique was spared further disaster.

Unlike in many MEDCs most Mozambique residents were too poor to take out insurance against flood losses. In addition, the government was too poor to cover the losses of the poor. Mozambique is a poor country (Table 9) and found it very difficult to repair the damage that had been done. The Mozambique government asked for $450 million to rebuild roads, homes and lives. In contrast, the UN appealed for $13 million of aid to help the 800,000 who were at risk of disease. This amounts to about $15 per head to cover health care, food and accommodation.

In the months following the floods the government started to regenerate the economy. For example, although the floods ruined 10 per cent of the country, destroyed up to 90 per cent of irrigated land, and washed away some 200,000 cattle, some Mozambique farmers returned to their lands and planted beans and maize in the silt. Crops sown before the southern winter (May to August) produced a quick

	UK	**Mozambique**
Population (millions)	58.1	18.0
Population density	238/km²	20/km²
Population growth rate	0.14%/year	2.6%/year
Population aged ≤ 15 years	19.3%	25.0%
Population aged ≥ 65 years	15.8%	4%
Male:female ratio	96.2:100	85:100 (effect of the civil war and labour migration to South Africa)
Human development index (HDI)	93.1	36
Life expectancy Male	74.5 years	44 years
Life expectancy Female	79.8 years	46 years
Literacy	99%	58% male 23% female
Fertility rate	1.7	6.1
Urban population	89%	35%
Crude birth rate	12‰	44‰
Crude death rate	11‰	18‰
Infant mortality rate	6‰	123‰
Employment structure		
Agriculture	2%	83%
Industry	27%	8%
Services	71%	9%
GNP	$19,600	$80

Table 9 The UK and Mozambique compared

harvest. The year 2001 was a relatively good year for Mozambique farmers as the silt deposited by the floods was rich in nitrates and phosphates.

In an attempt to guard against future floods, refugees were resettled on higher ground. In Xai-Xai aid agencies provided free building materials to people who helped clean up the flood debris from the town. In many cases, there was little left of them. Despite the devastation there was little unrest. There was evidence of some looting, and up to four people were killed in a stampede for food. Generally, however, the disaster was matched by patience and tolerance, unlikely virtues in a country that was at war with itself between 1975 and 1992.

What can never be quantified is the huge psychological trauma and distress that these floods have brought to the people of southern Africa. When the floods drained away, and the media departed, the people of southern Africa had to get on with their lives, and rebuild their countries. It is a tall order to ask, even of an MEDC. In Mozambique, where the average annual income is $80 a year (about £50 a year) such rebuilding is almost beyond comprehension.

Mozambique was again hit by flooding in March 2001 leaving up to 180,000 people living in temporary accommodation. The worst affected area was central Mozambique. The Zambezi River was 2–3m above its normal flood level for over five weeks, while continued rain upstream in Malawi and Zambia ensured that the river level remained high. Over 80 people died.

This time the Mozambique authorities were better prepared for the floods. Using equipment that was donated to them during the 2000 floods, they managed to begin evacuations early. The relief measures were the result of government efforts as well as non-government organisations (such as Oxfam) and international organisations, such as the UN World Food Programme and Unicef. About 80 tonnes of food each day were needed to support the 180,000 homeless people. However, food aid still took a long time to reach those in need, and the equipment was limited, two planes, two helicopters and a small number of trucks.

Flooding in the British Isles

From the middle of October 2000 to early April 2001 a large part of the British Isles was affected by the most widespread flooding in over 50 years. The result was that main roads and rail lines were blocked, many towns submerged and homes and businesses became waterlogged. Starting with severe flood warnings in East Sussex and Kent, the flooding continued to rivers in almost all parts of the country, particularly the Severn, with major floods in Shrewsbury and Bewdley, and the Ouse flooding parts of York. In York the Ouse reached flood levels that exceeded any for over 375 years.

A series of slow-moving low pressure systems generated intense rainfall over short periods of time, often over 50mm in a 24-hour period, repeated over several weeks. With the frontal systems circulating very slowly over most of the British Isles considerable quantities of rainfall were continually being added to already saturated ground and rivers at their bank-full discharge.

The scale of the downpour in such a short period made it impossible to prevent the flooding. Nothing short of brick walls around each town could have prevented it. Immediate defence schemes were needed to protect some of the 2 million homes in England and Wales built on floodplains in the 1970s and 1980s when planning was more relaxed. Most property at greatest risk was in the Thames and Trent catchments.

Plans for tens of thousands of new houses, particularly in the south-east and East Anglia, may have to be redrawn or scrapped because of poor flood defences. New planning rules will insist there must be 'sustainable drainage' so that water drains straight down into the ground rather than running along the surface.

Flood problems have been compounded by widespread intensive agriculture which has led to the drainage of wetlands in low-lying areas. The floodplains were a safety valve for flooding rivers. Now flood peaks are higher and occur in a much shorter time scale. The widespread cultivation of winter wheat sown in September leaves the land bare in the wettest part of the year. The run off of water from bare fields is 10–100 times greater than from a field with grass. The overstocking of sheep on the uplands has led to erosion resulting in more aeas of bare ground.

With flooding comes the risk of pollution. There are water borne health risks caused by spillages from oil tanks and contamination from sewers. Sewer systems become supercharged with floodwater which then spills onto the surface. Contamination problems were particularly acute in East Sussex where flooding put the main pumping station between Lewes and Newhaven out of use. Sewage pumping stations, chemical stores, oil stores and farms are all possible sources of pollution.

The Environment Agency has pinpointed a recent development at Eastbourne, known as Sovereign Harbour, as one of the most likely to be flooded or even swept away. The threat there comes not from rivers but from the sea. The marina and housing complex has 12,000 residents, who are protected by the flimsiest and most inadequate sea defences on the whole of the south coast.

4 High pressure and air quality

Poor air quality affects 50 per cent of the world's urban population, a total of about 1.6 billion people. Each year several hundred thousand die due to poor air quality, and many more are seriously affected. The problem is increasing due to increasing population growth in urban areas, industrial development and an increase in the number of vehicles worldwide. In Europe, three-quarters of cities with populations over 500,000 have poor air quality.

a) Air pollution and health

Air pollution is associated with high pressure. This is because winds in a high pressure system are usually weak, therefore pollutants remain in the area and are not dispersed. Air pollution has been linked with health problems for many decades. People most at risk include asthmatics, those with heart and lung disease, infants and pregnant women. This accounts for 20 per cent of the population in MEDCs

---

Asthma in the British Isles

The number of preschool children who wheeze doubled during the 1990s, worrying evidence of the steep rise in wheezing and asthma among children.

Researchers questioned the parents of 1650 preschool children living in Leicestershire in 1990, and the parents of a further 2600 under-fives in 1998. All types of wheezing had increased, not just the atopic kind (that arising from an allergic reaction) but the type which is classified as viral because of being linked to colds. Over the eight years the proportion of children reported to have had a wheeze rose from 16 to 29 per cent. Diagnoses of asthma rose from 11 to 19 per cent.

Some of the factors thought to account for wheezing, such as passive smoking, gas cooking and household pets, had declined and so could not be responsible for the increases. There are other social and environmental factors that have altered over time, e.g. diet and exercise patterns, but the cause of the increase remains unknown.

---

and an even higher proportion in LEDCs. The death rate from asthma has increased from 40 to 60 per cent in recent decades and it is now one of the most common causes of hospital admissions for children. The main pollutants which cause asthma are sulphur dioxide, ozone, acid aerosols, PM10s (very fine particulates which can flow into the lung), nitrogen dioxide and dust. Asthma is rising steadily and about one in seven children in the British Isles now suffer from asthma.

The Clean Air Acts 1956 and 1968 introduced controls on pollution by smoke, grit and dust from domestic and certain industrial sources, including controls on chimney heights and empowering local authorities to designate smoke control areas.

Since then there have been important changes in the source of pollutants, with motor vehicles an increasingly important source. Fuel combustion is the major source of sulphur dioxide (SO_2), oxides of nitrogen (NOx) carbon monoxide (CO), carbon dioxide (CO_2), hydrogen chloride (HCl) hydrocarbons and heavy metals. Industrial processes are also sources of these compounds.

b) Summer and winter smog

Poor air quality often persists for many days. This is because it is associated with stable high pressure conditions which generally prevail for a few days. In some climates, notably Mediterranean climates,

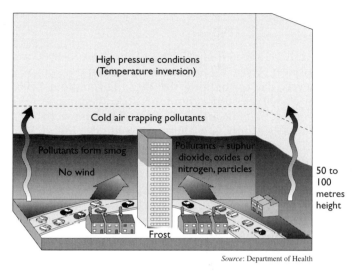

Source: Department of Health

Figure 26 The formation of winter smog

stable high pressure conditions persist all season, hence poor air quality can remain for months. In monsoonal areas smogs occur in the dry season. Although smogs occur under certain atmospheric conditions (namely high pressure), human activity (the emission of pollutants) is responsible for the environmental hazard.

Summer smog occurs on calm sunny days when photochemical activity leads to ozone formation. Ozone is formed when nitrogen oxides and VOCs react with sunlight. This process may take a number of hours to occur, by which time the air has drifted into surrounding suburban and rural areas. Hence ozone pollution may be greater outside the city centre. The effect of ozone pollution is to cause stinging eyes, coughing, headaches, chest pains, nausea and shortness of breath, even in fit people. Asthmatics may experience severe breathing problems. In city centres it is nitrogen dioxide (NO_2) rather than ozone that is likely to be the main pollutant. Vehicles emit two forms of nitrogen dioxide – nitric oxide and nitrogen dioxide. The nitric oxide is converted (oxidised) into nitrogen dioxide by reactions with oxygen and ozone. This in turn reduces the ozone concentration over city centres.

Winter smogs are associated with temperature inversions and high rates of sulphur dioxide and other pollutants, due to increased heating of homes, offices and industries (Figure 26). Under cold conditions, vehicles operate less efficiently and need more time to 'warm up'. This releases larger amounts of carbon monoxide and hydrocarbons. Urban areas surrounded by high ground are especially at risk from winter smogs. This is because cold air sinks in from the surrounding hills, reinforcing the inversion.

c) Forest fires

USA, 2000
The forest fires that affected large parts of the USA in 2000 were the worst there for over 50 years. The US government spent approximately £10 million every day on fighting the fires. Throughout 2000 there were well over 60,000 fires throughout the USA, affecting over 3.76 million acres, nearly double the annual average. Their impact was greatest in 11 states, notably Montana, California, Colorado, Idaho, Wyoming and Nevada. The fires occurred earlier than usual, and proved difficult to deal with. Some fires were so well established that firefighters knew it would take months to put them out. In fact only when the snows of October and November fell were the fires finally conquered.

The causes of the fires were both natural and human-made. Dry lightning was the main natural cause. Dry lightning is formed during periods of unusually high pressure and is associated with high temperatures, low humidities and rainless thunderstorms. Temperatures reached 40°C in many parts of the USA, making the land very susceptible to fires.

Human-related causes included discarded cigarette ends and barbecues that got out of control. Ecologically, it makes sense to prevent people moving into such vulnerable areas, but politically it is less easy to control population movements. The pressure on the forests of the USA is likely to increase in future rather than decrease,

Fire fighters used a number of techniques to fight the fires including bombing the fires with a mixture of slurry and water, as well as using fire to burn fire breaks in the vegetation. Ultimately, unless there is a low pressure storm with a large amount of rainfall, fire will remain a threat during times of high pressure.

Migration and hazards in dry areas

In temperatures over 35°C, you need to consume at least a gallon of water every hour. As a gallon weighs nearly 4kg, it is almost impossible for illegal migrants travelling from Mexico to the USA to carry enough water for their hazardous journey. Illegal migrants cannot carry enough water to last a day in the heat of the Arizona desert. Over 490 people died crossing the 3300km border into the USA in 2000. Up to 300,000 attempt to make the crossing every year, many planning their journeys as the summer demand for farm-workers grows. Arizona has become one of the most popular crossing points after the US government's 'Operation Gatekeeper' initiative in the 1990s made it harder to cross into California and Texas. The Mexican government has considered issuing survival kits, complete with rehydration powder and antidotes for snake and spider bites to those preparing for the crossing.

5 El Niño and La Niña

El Niño and La Niña are extreme phases of a naturally occurring climate cycle referred to as El Niño/Southern Oscillation. Both terms refer to large-scale changes in sea-surface temperature across the eastern tropical Pacific. Usually, sea-surface readings off South America's west coast range from the mid-teens to the mid-twenties C, while they exceed 28°C in the 'warm pool' located in the central and western Pacific. This warm pool expands to cover the tropics during El Niño, but during La Niña, the easterly trade winds strengthen and cold upwelling along the equator and the West coast of South America intensifies. Sea-surface temperatures along the equator can fall as much as 4°C below normal.

El Niño is a reversal in the atmospheric circulation in the eastern Pacific – it causes major changes in rainfall patterns, storms and drought. El Niño means 'Christ Child' because it originally occurred off the coast of Peru around Christmas time. During an El Niño event the normal east-west circulation in the Southern Pacific is reversed. The El Niño or 'little child' is the unusual warming of the sea surface in the western Pacific. This warm water current then moves eastwards towards South America disrupting the normal pattern of precipitation and air currents. The sea temperature rises to over 28°C – in some places the rise in temperature may be as great as 10°C. The El Niño has a specific pattern associated with it. For example, on the west Pacific side, in countries like the north eastern part of Australia, the Philippines and Indonesia, will expect much less rainfall than usual. At the other side of the Pacific, the eastern side, countries such as northern Peru, Ecuador and Bolivia are at a high risk of high rainfall and flooding. The results can be felt worldwide.

El Niño occurs anywhere between every two years and every ten years. The effects are global. The 1997–8 El Niño caused warmer conditions in eastern Canada/USA, drier conditions in eastern Brazil, and wetter conditions in Argentina.

El Niño have caused considerable damage – 2000 deaths and $13 billion worth of damage were blamed on the El Niño in 1982 and 1983 while the 1992–3 event was responsible for 160 deaths and caused over $2 billion of damage. The 1997–8 event led to storms in California, flooding in Southern USA, drought in Australia, the Philippines, north east Brazil and Southern Africa.

La Niña means 'The Little Sister.' La Niña is the opposite of El Niño and occurs at irregular intervals. La Niña is defined as cooler then normal sea-surface temperatures in the tropical Pacific ocean that impact global weather patterns. Satellite observations have revealed unusually cool ocean temperatures in the eastern equatorial Pacific. In the tropics, La Niña tends to cause the opposite of El Niño. In the United States, La Niña brings wetter than normal conditions across the Pacific Northwest and drier and warmer than normal con-

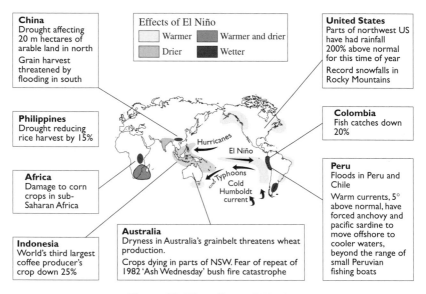

China
Drought affecting
20 m hectares of
arable land in north
Grain harvest
threatened by
flooding in south

Effects of El Niño
Warmer   Warmer and drier
Drier   Wetter

United States
Parts of northwest US
have had rainfall
200% above normal
for this time of year
Record snowfalls in
Rocky Mountains

Philippines
Drought reducing
rice harvest by 15%

Colombia
Fish catches down
20%

Africa
Damage to corn
crops in sub-
Saharan Africa

Hurricanes
El Niño
Typhoons
Cold
Humboldt
current

Peru
Floods in Peru and
Chile
Warm currents, 5°
above normal, have
forced anchovy and
pacific sardine to
move offshore to
cooler waters,
beyond the range of
small Peruvian
fishing boats

Indonesia
World's third largest
coffee producer's
crop down 25%

Australia
Dryness in Australia's grainbelt threatens wheat
production.
Crops dying in parts of NSW. Fear of repeat of
1982 'Ash Wednesday' bush fire catastrophe

Figure 27 The effects of El Niño

ditions across much of the south. At higher latitudes things are more complicated but in general, winters are warmer and summers are cooler. La Niña conditions typically last approximately 9–12 months. Some episodes may persist for as long as two years.

The scientific consensus is that El Niño will become more severe and possibly more frequent on a warmer planet, but it is not proven. There are things people can do, provided they are given enough warning. For example, Peruvians change the crops they plant – switching from maize to rice. They change the pattern for fishing from small boats in coastal areas to larger ones which can go into the ocean.

Better predictions of the potential for extreme climate episodes like floods and droughts could save billions of dollars in damage costs. Predicting the onset of a warm or cold phase is critical in helping water, energy and transportation managers, and farmers plan for, avoid or mitigate potential losses. Advances in improved climate predictions will also result in significantly enhanced economic opportunities, particularly for the national agriculture, fishing, forestry and energy sectors, as well as social benefits.

The chances for the continental U.S. and the Caribbean Islands to experience hurricane activity increases substantially during La Niña. El Niño and La Niña also influence tornado activity. Since a strong jet stream is an important ingredient for severe weather, the position of the jet stream determines the regions more likely to experience tornadoes. During El Niño the jet stream is oriented from west to east over the northern Gulf or Mexico and northern Florida. Thus this region is most susceptible to severe weather. During La Niña the jet stream extends from the central Rockies east-northeastward to the eastern Great Lakes. Thus severe weather is likely to be further north and west during La Niña than El Niño.

Summary

- Natural hazards endanger people's lives, properties and/or economic livelihood.
- Natural hazards appear to be having increasing impact on society. This may be due to improved reporting, more people at risk, or increasing high magnitude events.
- People continue to live in areas prone to hazards because the environment is perceived as being useful in some way.
- People cope with hazards in many ways – accept it, adapt to it, cope with it, ignore it, or leave the area.
- Hurricanes are tropical storms characterised by low pressure, intense winds and heavy downpours.
- The Mozambique floods of 2000 and 2001 wreaked havoc on a country emerging after years of civil war.
- In the UK floods appear to be getting more frequent and more intense.
- High pressure systems are associated with poor air quality which affects the lives of millions of people worldwide. High pressure conditions also help forest fires to develop.
- The El Niño and La Niña phenomena are major disruptions and intensifications of the global circulation model. They are linked with climate hazards around the world.

Questions

1. What are the causes, consequences and potential solutions to poor air quality in urban areas? (*15 marks*)
2. How does the hazard of poor air quality in urban areas differ between MEDCs and LEDCs? Use examples to support your answer. (*10 marks*)

Advice

1. At first glance, quite straightforward – there are three sections (causes, consequences and solutions) and 15 marks – so 3 × 5. However, it is possible that a very good answer on causes, for example, but weak on the other two sections, could score 7 + 1 + 1.

Causes should refer to factors such as industry, vehicles, domestic sources, high pressure conditions and types of pollutants. Examples will clearly help.

Consequences could be the number of days with poor air quality, the effect on health, acid precipitation, etc.

Solutions are varied including banning cars, use of catalytic convertors, Clean Air Acts, controls on industry, promotion of public transport. Again, named examples will help.

A mark scheme may look like this:

Level I   (0–2 marks)
Garbled account; no real detail or focus; a few rambling, unconnected points.

Level II   (3–5 marks)
Descriptive account; one or two points on each of the three areas; depth and/or detail lacking.

Level III   (6–8 marks)
Good description; uses examples; less detail on explanation; mostly causes and/or consequences.

Level IV   (9–11 marks)
Thorough account; detailed description and explanation; limited discussion of solutions; a few sweeping statements.

Level V   (12–15 marks)
Measured account of all aspects of question; balanced and evaluative; good use of case studies.

2.   Again, on the face of it a 2 × 5 LEDC and MEDC question.
LEDCs – greater impact on people, less impact on the economy
MEDC – greater economic loss, fewer lives lost.

In general, MEDCs have early warning systems and greater medical facilities available whereras LEDCs have fewer facilities and fewer early warning systems. However, pupils may take examples such as Los Angeles to show that MEDCs have poor air quality whereas cities such as Curitiba in Brazil have good air quality. Rapid urbanisation and industrialisation normally lead to poor air quality, however. A typical mark scheme may look as follows:

Level I   (0–2 marks)
Very vague description; sweeping comments regarding LEDCs and MEDCs; no real substance.

Level II   (3–4 marks)
Descriptive account; some idea of differences between LEDCs and MEDCs but no real depth/use of detailed case studies.

Level III   (5–6 marks)
Good descriptive account; details provided; limited explanation.

Level IV   (7–8 marks)
Good description and explanation; differences linked to reasons why.

Level V   (9–10 marks)
Thorough account using case studies to suggest why differences occur. May tackle question and argue that MEDCs are not better off than LEDCs, using evidence to support argument.

7 Weather, climate and society

Applied meteorology The study of weather and climate and its impact on human activities.

Biometeorology The study of the effects of weather and climate on people (and animals and plants).

Environmental determinism A philosophy popular in the early twentieth century that stated that if certain environmental conditions exist, the result can be predicted.

Possibilism A philosophy which states that humans can act in a variety of ways in a given environment.

Probabilism A philosophy which states that humans can act in a variety of ways in a given environment, but that some actions are more probable than others.

Experiencing the weather is probably most people's most direct and frequent experience of the physical environment. This is especially true for those who live in urban areas, and are less likely to experience other aspects of the physical environment. The impact on human activity ranges from the spectacular to the mundane. For example, the sinking of the *Titanic* might not have happened had there been better visibility and calmer seas. The weather affects what we wear, how we feel, and, in some cases, levels of health. There are many ways in which climate change impacts on human society. There are direct and indirect effects; short-term changes and long-term changes. Climatic stresses are physical, but the effects are social, economic, political and environmental.

1 Environmental determinism and probabilism

Biometeorology is the study of the effects of weather and climate on people (as well as animals and plants). In addition to the natural environment, it is possible to study the microenvironment of the home, school or office and its effects on people's well-being. An extreme view is that of *environmental determinism*. This states that if certain environmental conditions exist, the result can be predicted. According to Ellen Semple (1911) 'hot, moist equatorial climates encourage the growth of large forests which harbour abundant game and yield abundant fruits, they prolong the hunter gatherer stage of development and retard the advance to agriculture'. Such an extreme view has been criticised on two main counts. First, similar environments do not always produce the same result. The Greek and Roman empires flourished in Mediterranean climates, but there have not been similar empires in other Mediterranean areas such as California, South Africa and south-east Australia. Second, determinism fails to recognise the ways human activity can affect the environment.

Rather than disregarding the influence of the environment entirely, the idea of *possibilism* was developed. This suggests that humans can act in a variety of ways in a given environment. *Probabilism* is similar to this, but states that some actions are more likely than others. This gives humans choice, and makes them active agents within the environment, but it also suggests that the environment sets limits within which human activity must take place.

It is not easy to measure the influence of the environment on human activity but many have attempted to. Aristotle believed that people from cold climates were brave but lacked thought, whereas those in warm areas were thoughtful but lacked spirit. Not surprisingly, most writers concluded that their own environment brought out the best in people. For example, Huntington (1915) and Markham (1944) concluded that temperate areas with frequent changes in weather, such as in western Europe, north-east USA and Japan, stimulate mental activity and leadership.

There is evidence that climate and weather have an impact on a variety of features such as human comfort, patterns of disease, crime and suicide. Environments vary in terms of their suitability for human activity. Studies on twentieth-century white European populations have indicated that in still air, out of the Sun the average upper limit of 'comfortable' temperatures is about 22°C for 100 per cent relative humidity, rising to 27°C when the relative humidity is 66 per cent and 38–39°C with very dry air. The upper limit of what is 'just bearable' appears to go from 38°C with 100 per cent relative humidity to about 56°C in very dry air. Humidities over 90 per cent are found to induce feelings of lethargy whatever the temperature.

An important influence of weather and climate on human activity is though the effects of hazards. Climatic hazards such as tornadoes, hurricanes, storms and droughts are dealt with in Chapter 6. However, there are others, such as fog, poor driving conditions, wind chill and sunshine intensity, which all affect people.

Weather and climate have a direct impact on health and death rates. Death rates increase once a critical temperature has been reached. Mortality rates are higher in cloudy, damp, snowy places. Certain diseases have a seasonal pattern. Figure 28 shows the seasonal pattern of diseases in London and South Africa.

In addition, climate and weather have an important effect on economic activities (Table 10). The most obvious example is that of agriculture:

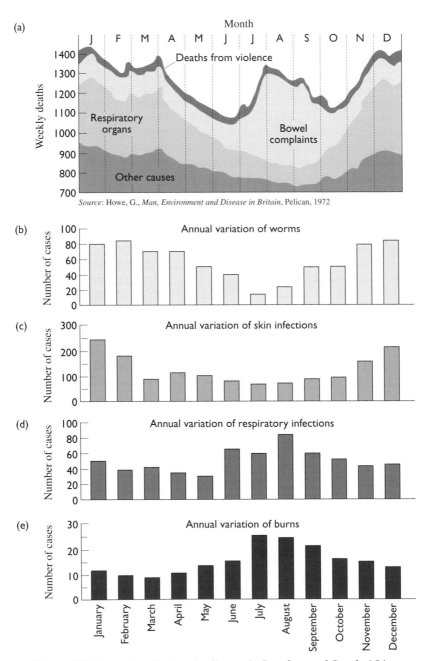

Figure 28 Seasonal variations in disease in London and South Africa

	General activities	Specific activities
Food	Agriculture	Land use, crop scheduling and operations, hazard control, productivity, livestock and irrigation, pests and diseases
	Fisheries	Management, operations, yield
Water	Water disasters	Flood, drought and pollution control
	Water resources	Engineering design, supply, operations
Health and community	Human biometeorology	Health, disease and mortality
	Human comfort	Settlement design, heating and ventilation, clothing, acclimatisation
	Air pollution	Potential, dispersion, control
	Tourism and recreation	Sites, facilities, equipment marketing, sports activities
Energy	Fossil fuels	Distribution, utilisation, conservation
	Renewable resources	Solar, wind and water power development
Industry and trade	Building and construction	Sites, design, performance, operations, safety
	Communications	Engineering design, construction
	Forestry	Regeneration, productivity, biological hazards, fire
	Transportation	Air, water and land facilities, scheduling, operations, safety
	Commerce	Plant operations, product design, storage of materials, sales planning, absenteeism, accidents
	Services	Finance, law, insurance, sales

(*Source*: Goudie, A (Ed.) 1994 *The encyclopaedic dictionary of physical geography*, Blackwell)

Table 10 Applied meteorology: sectors and activities where climate has significant social, economic and environmental significance

- plants require water to survive and grow – too much may cause soil erosion, too little may cause plants to die
- most plants require temperatures of over 6°C for successful germination of seeds; *accumulated temperatures* are the total amount of heat required to produce the optimum yield (wheat needs about 1300°C)
- wind can increase evapotranspiration rates and erode soil
- cloud cover may reduce light intensity and delay harvesting.

Increasingly, climate is having an impact on manufacturing activity. In the USA, states with warmer climates, such as Arizona and New Mexico, have a cost advantage over colder states. This is because firms need to spend less on heating bills, as do workers.

Climate also influences people, in the way that it varies. These variations include long-term changes (such as global warming), seasonal changes (such as decreased temperatures in winter) and rapid, unexpected changes (such as storms). As a result there is a great deal of human interest in being able to predict the weather, and not to be caught out.

2 Early climatic impact on humans

a) The beginnings of agriculture and the herding of animals

Until 10,000 BC the human inhabitants of Iran, Iraq, Turkey and Syria had been living in caves and hunting wild game, largely sheep and goats. Hunters lived mostly in the mountains because of the availability of wild game for food and caves for shelter. With the change to a warmer climate and the introduction of wild grains there was a combination of circumstances more favourable for domestication of animals and the growth of barley. According to Wright (1968) cultural evolution (the gradual refinement of tools and techniques for controlling the environment) is a stronger force than climatic determination in the development of early cultures. However, the coincidence of important environmental and cultural change in this area during the initial phases of domestication cannot be ignored.

The rise of Egypt and the organised cultivation of the Nile valley by use of the yearly flood for irrigation, may have been a necessary response to the drying of desert terrain in northern Africa at the same time. It was also made possible by the knowledge of agricultural techniques that must have been spreading in this region.

b) Early civilisations

According to Lamb (1971) just as civilisations in Egypt had developed, civilisations in the Tigris and Euphrates lowlands, in the Indus valley and in China were at least in part a necessary development to feed a more concentrated population at a time when huge areas outside those valleys in Arabia, Afghanistan, Rajasthan and the Gobi, were becoming more desert-like. If pastures and stocks of wild game were failing, the advantages of cultivation in irrigated valleys would be more attractive to those faced with abandoning an age-old way of life. It was the refugee herdsmen and farmers from the increasingly desert-like regions who became the slaves in such areas and made possible the intensive agriculture and the great building works for which ancient Egypt and the other river valley civilisations are famous.

A more serious break in the climatic regime came between 3500 and 3000 BC. This was the so-called *Piora Oscillation*, named after the Val Piora where the first evidence (by pollen analysis) was found indicating a cold episode. In the temperate forests of Europe and parts of North America the more warmth-demanding trees, the elms and the linden (or so-called lime) trees, declined and never regained their former, dominant, position in the forests. It is not certain whether, or to what extent, human interference or browsing cattle played a part in this, but the phenomenon seems to have been too widespread for this to be the main explanation. For a time in northern Europe the oak declined too, and the hazel withdrew for good from its northern limits. The duration of this colder episode seems to have been quite short, at most four centuries.

In Europe around 3000–2000 BC the climate warmed again and forest and grassland limits extended northwards beyond their present-day positions. In the north and in the continental interiors the permafrost was more restricted than at present and glaciers were less advanced. In Iceland some valleys that are now filled with ice supported trees. A measure of the relative warmth of those times is found in the relics which established the tree line, and also in the extension of Bronze Age cultivation on the hills of Dartmoor to over 450m above sea level, compared with the absolute limit of 300m in the same district today.

In southern England, by 2000 BC, human activity had been considerable, and was disturbing the natural vegetation cover, particularly near the chalk uplands. The hilltops were abandoned except for grazing and burials and for their convenience as travel routes.

Between about 800 and 400 BC a very wet period in the west caused the great bog at Tregoran in Wales to grow by nearly one metre. In contrast, in eastern England conditions became much drier. Meteorological reconstruction suggests that a cooling Arctic had pushed the cyclonic activity south over northern Europe. The period was associated with some outstanding storminess. About 500 BC the climate became much wetter than before in eastern districts of England; thus all parts of England, Wales and Ireland were then affected by the notably wet regime.

c) The Mediterranean world in Roman times

Roman horticultural writers drew attention to the fact that the vine and the olive could be cultivated further north in Italy than in earlier centuries. This suggests warming in Europe through Roman times, and increasing dryness, until about AD 400.

The extensive African croplands provided food for the Roman empire, and supported thriving settlements such as Petra that have since been conquered by the desert. The weather diary kept by Ptolemy of Alexandria (AD 120) shows some remarkable differences from today's climate in that there is the occurrence of rain in every month of the year except August.

During Roman times, from about 150 BC until AD 300, caravans of camels used the Great Silk Road across Asia to trade luxuries from China. However, by the fourth century AD, drought developed on such a scale as to stop the traffic along this route. Other serious stages of the drought occurred between abut AD 300 and 800.

d) Viking times and the Middle Ages

By the late tenth to twelfth centuries most of the world seems to have been enjoying a renewal of warmth. The northern limits of the cultivation of grains show the following expansions:

- grain was grown in Iceland
- in Norway barley was grown as far north as 69°N
- in many parts of the British Isles tillage was extended to greater heights than for a long time previously or since: on Dartmoor in the south-west to about 400m and in Northumberland, near the Scottish border to 320m
- some of the problems with cathedral towers collapsing were due to soil moisture changes and consequent settling
- the occurrence in mediaeval York of the bug *Heterogaster urticae* whose typical habitat today is on stinging nettles in sunny locations in the south of England, indicates prevailing temperatures higher than today's.

e) Cooling and wetness in early fourteenth-century Europe

A cooling trend began to affect Europe soon after 1300. In 1315, the grain failed to ripen across Europe, and famine in many parts of the continent was so bad that deaths from hunger and disease occurred on a great scale, and cannibalism was reported. Large numbers of sheep and cattle died in epidemics which swept the flooded landscape. The prevailing wetness during parts of the fourteenth and fifteenth century made this an unhealthy time. During the 'Black Death' which arrived in 1348–50 between one-eighth to two-thirds of the population died. Desertion of farms and village settlements was

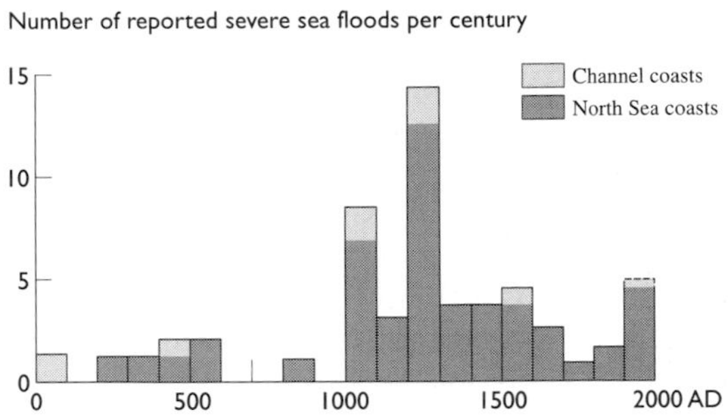

Number of reported severe sea floods per century

Figure 29 The distribution of severe storms on the Channel and North
Sea coasts

widespread. The fact that climatic change played a part can be seen
in:

- the failures of the northern vineyards in England
- the retreat of corn-growing from its former northern limits
- the depopulation of villages and farms.

Explorations in the fifteenth century which led fishermen from
Bristol ever further west across the Atlantic may have started because
the fish stocks of the higher latitudes in the north-east Atlantic had
deserted their former grounds as a result of the increasing spread of
the Arctic cold water. There were increased incidences of wind storms
and sea floods in the thirteenth century; some caused appalling loss
of life (Figure 29). In at least four sea floods of the Dutch and
German coasts in the thirteenth century the death toll was estimated
at around 100,000 or more; in the worst case the estimate was
306,000.

f) The Little Ice Age

In the middle of the sixteenth century a dramatic change in the climate of the British Isles occurred. The next 150 years experienced the
coldest regime (accompanied by great variations from year to year) to
have occurred at any time since the last major ice age ended 10,000
years previously. The whole period between about 1420 up to 1710 is
referred to as the Little Ice Age. One of the most long lasting impacts
of this era is that Scots farmers moved to the richer lowlands and more
sheltered climate of east Ulster in north-east Ireland after first evict-
ing the native Irish. This appears to have been a device of King James
VI to stabilise the Irish political and religious situation in his favour
and to relieve the impact of harvest failures in Scotland. By 1691 there
were 100,000 Scots in Ulster, already about a tenth of the population
of Scotland. Between 1693 and 1700 the harvests (largely oats) failed
in seven years out of eight in all the upland parishes of Scotland.

The long-term average winter temperatures for central England between 1670 and 1700 suggest that the normal yearly number of days with snow lying must have been 20–30 (compared with 2–10 days which has characterised much of the present century). The lowered summer temperatures in and around the 1690s were probably more important economically than the severity of the winters. In England the growing season was shortened by about five weeks in comparison with the warmest decades of the twentieth century.

One of the most remarkable responses to the climatic stress of the climax of the Little Ice Age was in southern Norway around the coast between Trondheim and Oslofjord. In the late seventeenth century, when harvests were poor and the grain sometimes failed to ripen on the farms even in the most favoured areas, farmers took to trading timber, notably to England. Those near enough to the coast built their own ships to carry it in. This seems to have been the beginning of what became two of Norway's greatest industries, the timber trade and her merchant fleet. Hence the years 1680–1709, the bitterest period of the climate in northern Europe, are described as 'the first great period of Norwegian shipping'.

g) The Recovery 1710–1950

Temperature records show a sharp change to much warmer conditions in Europe in the 1730s which produced a run of warmth equalling the warmest part of the present century. Even in this warming period there were several more cold winters, whereas the summers of 1718 and 1719 produced great heat and drought over most of Europe. However, 1725 produced the coldest summer with a mean temperature over June, July and August of just 13.1°C, and the decade 1810–19 was the coldest in England since the 1690s.

The summer of 1846, which was warm in Europe generally, was humid and cyclonic, with moist southerly winds. This provided ideal conditions for the potato blight fungus (*Phytophthora infestans*) which had made its first appearance in Europe in 1845 and spread quickly. *Phytophthora infestans* multiplies rapidly when temperatures are continuously above 10°C and relative humidity remains above 90 per cent saturation. In 1846 the potatoes rotted. In Ireland, where the potato was the staple crop on the multitudes of small farms, the effect was devastating. Despite relief measures, upwards of 1 million people died over a six-year period. The population in 1851 had already dropped by nearly a quarter from its peak of 8 million in 1845, and by the twentieth century it had fallen by a half and has never since approached the 1845 level.

The summer of 1868 produced a remarkable number of hot days with temperatures over 30°C in England, including the record value of 38.1°C at Tonbridge, Kent on 22 July. During the 1870s Europe enjoyed mostly warm summers and mild winters. However, between December 1878 and January 1879 the temperature in England stayed below freezing point. Spring was cold and the summer was wet and cold; it was followed by a cold autumn and another near freezing winter. The cold, wet weather delayed the ripening of the harvest, so that even in East Anglia the corn had not been gathered in by Christmas. The 50-year decline of English agriculture dated from this time:

*   harvests had been affected by difficult seasons from 1875
*   competition with cheap North American wheat from the prairies was beginning to be felt
*   within a few years the cornlands of the north-west of England had been converted to grass, and brought no profit
*   meat began to come from Australia, New Zealand and South America
*   farmworkers began to leave the land for the towns and to emigrate overseas in great numbers
*   other European countries protected their farmers against the American competition by import dues
*   the peak emigration of people from Europe was in the 1880s.

3 Climate in the twentieth century

In the USA in the early twentieth century the increasing scale of mechanised farming operations, taking in more and more of the great grasslands, gradually caused earlier droughts to be forgotten, until disaster struck in the 1930s. Successive summers between 1932 and 1937 brought hot, dry winds from the Rockies which parched both vegetation and soil in the prairie lands. Thousands of farmers were ruined in those infamous years when the Midwest became a 'Dust Bowl', many families migrated to seek a new living near the west coast, and farms inland were abandoned. Soil rehabilitation programmes had to be instituted by the government, involving returning much of the land to pasture and planting trees as windbreaks.

A similar mistake was made in the Sahel in the time of more abundant rains the 1950s and early 1960s. International aid for LEDCs in the region was used to drill deep wells in order to use (and ultimately use up) the great reserves of subterranean water which had accumulated in different climatic regimes for thousands of years. This introduced short-lived prosperity to the region with greatly increased cattle herds and growth of the human population. However, the sparse vegetation was soon over-grazed, resulting in a spread of the desert and resulting in subsequent droughts. Drought was caused in this way for the following reasons:

*   through the greater reflection of the solar radiation by bare soil the total energy absorbed in the ground and lower atmosphere is reduced and an anticyclonic tendency with dry air subsiding from aloft is introduced
*   at the same time there would be even less moisture than before stored in vegetation and available for recycling.

In these ways the whole region became more vulnerable to the next shortfall of rainfall, which occurred in the mid-1960s onwards. There seems to be a sort of historical cycle, whereby human populations expand in periods of benign climate and occupy with increasing density lands which sooner or later fail to support the numbers by then dwelling in them. Similar expansions of population are introduced by advances in technology. When the bad years come, the population has always in the past been reduced or disappeared, partly through migration and partly through undernourishment, disease and death.

The doubling of wheat and corn yields in the USA between 1955 and 1973 was achieved by technological innovations and a long run of drought-free years. Similarly agriculture in western Europe gained from the warm period between 1933 and the 1950s. The advances in agricultural development are now threatened on a world scale by the growth of population. Since the best land for agriculture is already in use, one must expect lower returns from any further increases of the area sown. Traditionally the population of drought-prone areas guarded against disaster by planting a wide variety of crops, so that at least some were likely to survive any drought periods. For the same reason they kept great numbers of cattle that could roam over an extensive area. The increase of population and modern agricultural development since 1950 have made these safeguards impossible. The development of monoculture on a world scale, with concentration on just one or two crops in each region where they are supposed to grow best, constitutes a threat to this security. Monoculture was at the root of most of the great famines of the past. The selection of areas where each crop grows best implies a forecast of no climatic change. This seems to have been an important aspect of the Irish potato famine of the 1840s. The success of the potato in the moist climate of the Atlantic fringe of Europe, and the growth of population in Ireland in the eighteenth and early nineteenth centuries, had meant that this was the one crop which could produce enough food for a family on the very small farms, many of which were less than one hectare. When the previously unknown blight appeared, and was quickly spread by the winds of the autumn of 1845 and the moist summer of 1846, the impact was devastating.

For example, cooling in the Arctic has been severe. Similarly, cold years in Iceland can be disastrous since lower summer temperatures limit farm production. In the 1950s the mean summer temperature in Iceland was 7.7°C and average hay yields were 4.3 tonnes/hectare. In the late 1960s, with mean temperature 6.8°C, the average yield was only 3.0 tonnes/hectare. This temperature level is close to the point at which grass ceases to grow.

Climatic variability has produced some great droughts in the former Soviet Union. 1972 was one of the most severe years. The overall average rainfall declined; a huge area of central Asia had under half the usual rainfall, and in most of that area the totals were under a quarter of the expected amount. With the average temperature of the summer up to 3.7°C above the long-term averages, the great drought of the former Soviet Union ruined crops, caused extensive forest fires and even set the dried-up peat bogs on fire. The government had to make massive wheat purchases from the West and it coincided with monsoon failures in India and West Africa. Food shortages were widespread and had a major impact on world trade.

In 1972, heat and drought in the former Soviet Union led to the grain harvest being 13 per cent below expectation. Similarly the Chinese harvest was disastrously poor and in northern India the monsoon failed. An estimated 100,000 to 200,000 people and 4 million cattle died between the Sahel and Ethiopia. The coffee harvest in Ethiopia, Kenya and the Ivory Coast and the ground nuts, sorghum and rice in Nigeria were also sharply reduced. The Australian wheat crop in 1972–3, owing to drought there, was more than 25 per cent below the previous five-year average. The net effect was that the world's total food production in 1972 fell nearly 2 per cent below the 1971 achievement. This was the first drop that had occurred in any year since 1945. The world price of wheat doubled within a few months and the difficulties increased for the poorest countries suffering a shortage. The hopes that had been raised by the Green Revolution of being able to meet indefinitely the food demands of the world's rising population were seen to have been unduly optimistic, particularly since the high-yielding new varieties of rice and other 'wonder crops' were often more sensitive than the traditional varieties to deviations from the expected climatic conditions.

Summary

- The environment offers many possibilities and many constraints for human activities. It affects everyday life in many ways; some direct, many indirect.
- Climate changes over the millennia have had a great impact on civilisations.
- Agricultural change, migration and revolutions have all been linked to climate change.
- Major events in Europe – such as the Viking invasions and the Black Death have been attributed, in part, to climate change.
- Climate change has often worked through agriculture to produce major disasters, such as the Irish potato famine in the 1840s and the Dust Bowl in the 1930s.
- Climate change continued throughout the twentieth century and has a major impact on food supplies.

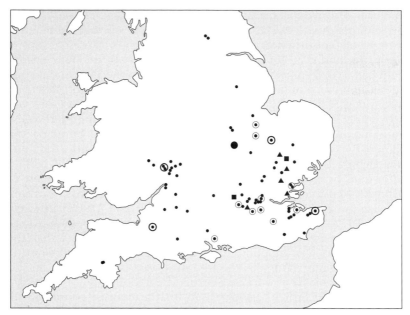

Key

- • Vineyard, usually 1–2 acres or size not known
- ▲ Vineyard, 5–10 acres
- ■ Vineyard, over 10 acres
- ◉ Denotes evidence of continuous operation for 30–100 years
- ◉ Denotes evidence of continuous operation for over 100 years

Figure 30 The distribution of medieval vineyard sites in England

Questions

1. Figure 30 shows the distribution of vineyards in medieval England.
   a) Describe and suggest reasons for the distribution of vineyards.
   b) How has the distribution changed over time?
2. With the use of examples explain how climate has affected human society.

Advice

2. A very general question requiring both a general and specific answer. The general answer requires an explanation of how climate affects society, for example:
   - water supply
   - temperature
   - sunshine
   - wind.

These have an impact on many activities including, for example:

1. agriculture
2. forestry
3. disease
4. health
5. manufacturing industry
6. trade
7. transport
8. tourism
9. disasters
10. insurance cover.

The question asks for examples, thus named and located studies are required.

Level I   0–4 marks
Rambling account; no real focus, a few valid points but unstructured or unlinked.

Level II   5–9 marks
Descriptive account; some fair points, with some attempt to order/classify.

Level III   10–14 marks
Accurate description; nominal/fair use of examples; partial explanation.

Level IV   15–19 marks
Detailed description; good use of case studies; fair attempt at explanation.

Level V   20–25 marks
Detailed description, probably containing sketch maps; case studies detailed and compared (evaluated); clear and successful attempt at explanation.

8 Contemporary climate change

In Chapter 7 we saw how environmental change, in particular climate-induced change, led to physical, economic, social and political changes. Many of these were long lasting and profound. In this chapter we look at the present and the foreseeable future. What is happening to the world's climate? How might climate change? What impacts might it have? What can be done about it? Whether the changes are short term or long term, natural or human-made, reversible or irreversible is open to question, but the changes themselves and their impacts are increasingly clear and alarming.

Approximately 1 billion people live at sea level or just a few metres above it. For them, the present and future effect of sea level rise caused by global climate change is potentially disastrous in terms of loss of life, home, land and livelihood. Computer models developed by the Intergovernmental Panel on Climate Change (IPCC) indicate that a 1m rise in sea level could occur by 2080.

1 Population at risk

Regional studies conducted by the IPCC suggest that the impacts of climate change will be drastic, especially in the tropics and warm temperate regions, where many coastlines are heavily settled. The most vulnerable areas include the southern coast of the Mediterranean, the west coast of Africa, South and South East Asia, and low-lying coral atolls in the Pacific and Indian Oceans. These regions contain some of the poorest and most heavily populated countries in the world; parts of China and South East Asia include areas with population densities of over 2000 people per square kilometre.

In Bangladesh, a 1m rise in sea level would flood 3 million hectares, affecting 15–20 million people. In India, 600,000 hectares would be drowned, and 7 million people displaced. Indonesia would lose 3.4 million hectares, home to at least 2 million people. Vietnam would lose 2 million hectares in the Mekong Delta and another 500,000 hectares in the Red River Delta displacing 10 million people.

The Maldives is an island nation of coral atolls off the south coast of India. A 1m rise in sea level would swamp about 85 per cent of the country's capital island, Malé, which has 60,000 people on just 600 hectares of land. Malé is on average just 1m above sea level. Nearly all of the islands would be converted into sandbars and tiny spits by an extra 1m of water, or they would sink beneath the waves completely.

In West Africa, countries along the Gulf of Guinea are facing a similar fate. In Nigeria up to 70 per cent of the coast would be inundated by a 1m rise, affecting more than 2.7 million hectares and pushing some beaches 3km inland. About 4 million people would become homeless and part of Lagos would be under water. Oil production in the Niger Delta would be disrupted. This might seem like an appropriate form of economic loss, given the cause of the sea level rise, but it is likely to be painful for the Nigerian economy all the same.

Egypt would be particularly affected by a 1m rise in sea level. At least 2 million hectares of the Nile Delta would be lost to the rising sea levels and a further 10,000 hectares of good farmland would be affected by erosion and salinisation. Up to 10 million people would be displaced, including most of the population of Alexandria. Most of the inundated area would consist of prime agricultural land currently worth close to $1 billion, but the loss of Alexandria would cost the country over $32 billion in lost land, infrastructure and tourist revenue.

a) The impact on urban areas

Rising sea levels provide a major problem for urban planners. In many coastal cities, the problem is exacerbated since land underneath the city is sinking. Excessive groundwater pumping is the main cause of this subsidence, but urban sprawl is important too. Buildings and pavements cause rainfall to run off instead of seeping back into the earth to recharge the groundwater. Underground saltwater intrusion is a serious problem for Manila, Dhaka, Bangkok and Jakarta. Most of Manila's wells, for example, will become too saline to use if the sea level rises by 1m. Another major expense is the need to improve flood control systems in the country. Manila's system is so old that every year during the monsoon rains, dozens of people drown in low-lying areas because the storm drains cannot handle the tremendous volume of water dumped on the city over the course of a few hours.

In the USA, the US Federal Emergency Management Agency (FEMA) estimated that a 0.5m sea level rise would inundate up to 1.9 million hectares of dry land along the eastern seaboard and Gulf Coast if no protection measures are taken. The FEMA study also found that a 1m rise would increase the amount of the east coast floodplain that is vulnerable to storm damage by 40 per cent. Over much of the floodplain, the frequency of storm damage would increase dramatically. Insurance costs would be beyond the reach of many people. Overall, according to the Environment Protection Agency study, a 1m rise in sea level could cost the US economy anywhere from $40 billion to $475 billion.

b) The impact on rural areas

According to the Hadley Centre for Climate Prediction and Research in Britain, 40–50 per cent of the world's remaining coastal wetlands will be lost by 2080, due to a combination of drainage for agriculture, urban sprawl and the effects of a 1m sea level rise. The coastal wetlands most likely to suffer lie along the Atlantic coast of North and Central America, the US Gulf Coast, and around the Mediterranean and Baltic Seas. There is little potential in such places for coastal wetlands to migrate inland, they would be trapped between the advancing waves and high ground. Many of these ecosystems will become far less diverse and probably less stable.

For example, the Sunderbans is the largest continuous mangrove forest in the world, covering 100,000 hectares along the Bay of Bengal, partly in India and partly in Bangladesh. It is ecologically diverse, including 315 species of birds. Among the endangered species are the Rhesus macaque (a monkey), the Irrawaddy dolphin and the Bengal tiger. A 1m sea level rise could lead to extinction for the local populations of many Sunderbans creatures. There are only about 350 Sunderbans tigers and these are an important reservoir of genetic wealth for their species. The rising waters would probably do away with their main prey and drive the tigers into heavily settled areas further inland, where it is unlikely that they would survive.

The Sunderbans is a human environment too. At least half a million people are directly dependent on the forest: woodcutters, thatch harvesters, fishers, and collectors of honey and beeswax. In the event of a 1m sea level rise, most of these people would share the tigers' fate; they would be forced further inland and their forest economy would be badly affected.

Similarly the Sahara desert is spreading northwards, forcing thousands to migrate as the combination of soil degradation and climate change turns parts of southern Europe into desert. Up to one-third of Europe's soil could eventually be affected. One-fifth of Spanish land is so degraded that it is turning into desert, and in Italy tracts of land in the south are now abandoned and technically desert. Portugal, Spain, Italy and Greece are the four EU countries already so badly affected that they have joined the United Nations Convention to Combat Desertification (CCD). Land that has been carefully cultivated and preserved for 2000 years, with terracing for soil conservation and careful irrigation to keep up productivity, is being abandoned and lost. Once the walls of the terraces break down, the soil is washed away leaving bare rock. The conditions are particularly bad in southern Italy, Spain and Greece. The problem is not confined to the EU; Bulgaria, Hungary, Moldova, Romania and Russia have all reported signs of desertification. Experts say Moldova in particular is 'highly vulnerable' to desertification, with about 60 per cent of its farmland degraded. Beyond the Black Sea, there are belts of fast-degrading land stretching as far as Mongolia. In China, for example, land deterioration in its northern provinces is costing its economy £4 billion a year.

Although often overlooked, soil is a natural resource that is no less important to human well-being and the environment than clean water and clean air. The sustainable use of soil is one of Europe's greatest environmental, social and economic challenges. In some parts of Europe, the degradation is so severe that it has reduced the soil's capacity to support human communities and ecosystems and resulted in desertification. Because it can take hundreds or thousands of years to regenerate most soils, the damage occurring today is, for all purposes, irreversible.

At the 2000 Convention to Combat Desertification, countries in southern Europe produced action plans to deal with:

* seasonal droughts, very high rainfall variability and sudden high intensity rainfall
* poor, highly erodable soil, prone to surface crusts
* crisis conditions in traditional agriculture with associated land abandonment
* an increase in forest fires
* concentration of economic activity in coastal areas.

2 The greenhouse effect and global warming

There are a number of reasons why the Earth's temperature changes, one of the most obvious being a change in the output of energy from the Sun. Slow variations in the Earth's orbit affect the seasonal and latitudinal distribution of solar radiation and these are responsible for initiating ice ages. On a shorter time scale, changes in atmospheric composition are linked with an increase in global temperature. The Earth's atmosphere is vital for life, and changes to it disrupt the natural balance of the Earth's energy budget.

Solar radiation is radiated mostly in the visible wave band between 0.4 and 0.7 micrometres (Figure 31). This radiation and short wave infrared radiation, passes through the atmosphere without being absorbed, (although clouds may reflect some of it) and most of it reaches the Earth's surface and warms the land and the sea.

Water vapour in the atmosphere absorbs radiation in the 4–7 micrometres band and carbon dioxide absorbs radiation in the 13–19 micrometres band. Between 7 and 13 there is a 'window' through which more than 70 per cent of the radiation from the Earth escapes into space. Roughly 7 per cent of solar energy is radiated at shorter wavelengths, below 0.5 micrometres. This is ultraviolet radiation and is important in maintaining a layer of ozone in the atmosphere. The infrared heat that is re-radiated from the Earth warms the lower part of the atmosphere (troposphere). In turn air in the troposphere radiates heat back towards the ground and this is known as the

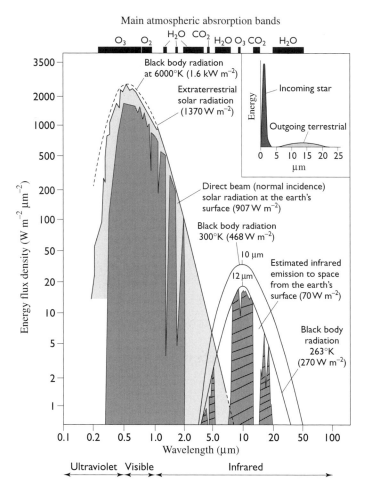

Spectral distribution of solar and terrestrial radiation, plotted logarithmically, together with the main atmospheric absorption bands. The hatched areas in the infrared spectrum indicate the 'atmospheric windows', where radiation escapes to space. The black-body radiation at 6000 K is that proportion of the flux which would be incident on the top of the atmosphere. The inset shows the same curves for incoming and outgoing radiation with the wavelength plotted arithmetically on an arbitrary vertical scale.

Source: Barry, R., and Chorley, R., *Atmosphere, Weather and Climate*, Routledge, 1998

Figure 31 Incoming solar radiation

Greenhouse Effect. Both the ground and the air above it are warmed by the Greenhouse Effect. As long as the amount of water vapour and carbon dioxide stay the same and the amount of solar energy remains the same, the temperature of the Earth should remain in equilibrium. However, human activities are upsetting the natural balance by increasing the amount of carbon dioxide in the atmosphere, as well as the other greenhouse gases.

There are a number of greenhouse gases, such as water vapour, methane, ozone, nitrous oxides and chlorofluorocarbons (CFCs). These mainly absorb infrared radiation in the 7–13 micrometres band where radiation has been able to escape freely. The best known are the CFCs and these are held to be responsible for the ozone hole over Antarctica. One molecule of CFC has the same greenhouse impact as 10,000 molecules of carbon dioxide. Methane is another greenhouse gas, at present it has an atmospheric concentration of about 1.7 ppm and is increasing at a rate of about 1.2 per cent per annum. This is largely due to the biological activity of bacteria in paddy fields and also due to the release of gas from oil and gas fields. Due to the increase of nitrogen based fertilisers the amount of nitrous oxide (NOx) is increasing from a concentration of 0.3 ppm at an annual rate of 0.3 per cent. Ozone near the ground in the troposphere is also increasing as a result of human activities. By 2030 increases in these minor greenhouse gases will probably have the same impact as the doubling of CO_2 from 270 ppm to 540 ppm.

Accurate measurements of the levels of CO_2 in the atmosphere began in 1957 in Hawaii. The site chosen was far away from major

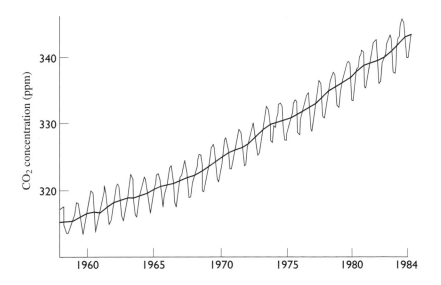

The build-up of carbon dioxide in the air, recorded at Mauna Loa Observatory, Hawaii

Source: New Scientist, 22 October 1988

Figure 32 The build-up of carbon dioxide in the air, recorded at Mauna Loa Observatory, Hawaii

sources of industrial pollution and shows a good representation of unpolluted atmosphere. The trend in CO_2 levels shows a clear pattern, one of a long-term increase, superimposed upon the annual trends (Figure 32).

Studies of cores taken from ice packs in Antarctica and Greenland show that the level of CO_2 between 10,000 years ago and the mid nineteenth century was stable at about 270 ppm. By 1957 the concentration of CO_2 in the atmosphere was 315 ppm and it has since risen to about 360 ppm. Most of the extra CO_2 has come from the burning of fossil fuels, especially coal, although some of the increase may be due to the disruption of the rain forests. For every tonne of carbon burned, 4 tonnes of CO_2 are released. By the early 1980s 5 gigatonnes (5000 million tonnes) of fuel were being burned every year. Roughly half the CO_2 produced is absorbed by natural sinks, such as vegetation and plankton in the oceans.

Since the industrial revolution the combustion of fossil fuels and deforestation have led to a 26 per cent increase in carbon dioxide concentration in the atmosphere. In addition methane concentrations have more than doubled because of rice production, cattle rearing, biomass burning, coal mining and ventilation of natural gas. Nitrous oxide has also increased by about 8 per cent since preindustrial times, presumably due to human activities. The effect of ozone on climate is strongest in the upper troposphere and lower stratosphere.

If carbon dioxide emissions were maintained at 1995 levels they would lead to a constant rate of increase in atmospheric concentration of CO_2 reaching twice the preindustrial concentration by the end of the twenty-first century.

a) Some effects of the rise in greenhouse gases

Researchers have considered the effect of a doubling of CO_2 from the base line value of 270 ppm to 540 ppm. Such a rise would lead to:

- an increase of temperatures by about 2°C
- increased warming at the poles rather than at the equator
- changes in prevailing winds
- changes in precipitation
- continental areas becoming drier
- sea levels rising by as much as 60cm
- ice caps changing in size (they may increase due to more evaporation in lower latitudes and increased snowfall at higher latitudes).

Temperature

Global temperatures have been increasing over the past century due to the effect of greenhouse gases. However, other factors also affect global temperatures such as changes in ocean circulation, solar output, volcanoes and air pollution products such as aerosols. The global temperature change observed over the past 100 to 130 years of

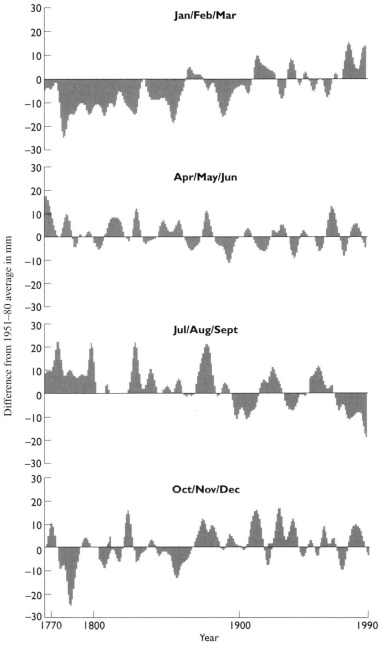

Notes: 1. Smoothed data. Monthly averages for each quarter.
Source: Gregory et al. (1991), Hadley Centre

Figure 33 Seasonal long-term changes in the distribution of rainfall

0.45°C is consistent with the expected increase in temperature estimated to result from increasing greenhouse gases, taking into
account the negative effect of ozone depletion and aerosols.
Although the recent warming over England is probably a reflection of
global changes, other fluctuations are known, in part, to be due to
changes in atmospheric circulation, possibly related to changes in sea
surface temperature.

Rainfall
There have been long-term changes in the seasonal distribution of
rainfall. Figure 33 provides evidence of a drying tendency in summer
and an increase in rainfall in winter. Again, changes in atmospheric
circulation are likely direct causes of these changes but the reasons
for the circulation variations are not known.

Sea level variations
Sea levels vary across a broad range of time and space scales for many
reasons. These include:

• long-term changes in ocean base and volume from, for example, sea
  floor spreading or sedimentation
• medium-term changes in ocean mass from variations in groundwater,
  surface water or land-based ice
• short-term dynamic changes due to oceanographic (e.g. currents) or
  meteorological (e.g. atmospheric pressure) factors at the local or
  regional scale.

As a result of global warming, mean sea level may change for two
reasons:

• expansion of the ocean due to higher sea temperatures
• changes in land-based ice.

The most recent projections suggest a rise of less than 1m before the
end of 2100.

Large-scale coastal movements can be important regionally. In
parts of Scandinavia and Hudson's Bay, relative sea level is rising
by about 1m per century. Elsewhere however, such as along the
east coast of North America and the south-east of Britain, sea level
is still rising due to land subsidence as a result of glacial isostatic
change. In other areas it is a very complex pattern, for example
the Mississippi Delta relative sea level change is determined by the
difference between the subsidence of the Delta, changes in sea
level and growth of the Delta due to increased deposition as a
result of river erosion. Rising sea levels are also threatening one of
most densely populated but poorest countries in the world,
Bangladesh.

b) Natural climate change

Not all climate change is due to human activity. Natural events, such as volcanic eruptions, can have an impact on climate change. For example, the volcanic plume associated with the Pinatubo eruption reached an altitude of more than 30km and injected 30×10^{12} G (30Tg) of aerosols into the atmosphere. The eruption of Pinatubo caused the largest injection of aerosols into the atmosphere in the twentieth century, but was much smaller than the eruption of Tambora in 1815 (over 100Tg) and Krakatoa in 1883 (approximately 50Tg).

Following the Pinatubo eruption there was a rapid movement of volcanic aerosols across the equator to about 10° South, it then moved to the band from 20° South to 30° North. Material erupted from volcanoes may remain in the stratosphere for many years. The injected material may include ash which typically does not remain for more than a few months, and gaseous components including water vapour, sulphur dioxide and hydrochloric acid. The new aerosol content in the atmosphere increases the Earth's albedo by reflecting solar radiation back to space.

3 Climate change in the British Isles

Between 1985 and 1994 temperatures in the British Isles were about 0.2°C warmer than between 1961 and 1990 and the average global atmospheric CO_2 concentration has risen by about 5 per cent. The government review of changing climate and changing sea level in the British Isles (1996) made the following predictions:

- temperatures are expected to increase at a rate of about 0.2°C per decade with slightly slower rates of increase over north-west Britain compared to the south-east, and in winter compared to the summer
- it will be about 0.9°C warmer than the average of 1961–90 by the 2020s and about 1.6°C warmer by the 2050s
- annual precipitation over the British Isles as a whole is expected to increase by about 5 per cent by the 2020s and by nearly 10 per cent by the 2050s
- winter precipitation will increase everywhere but more substantially over southern Britain
- sea level is expected to rise at a rate of about 5cm per decade – this is likely to be increased in southern and eastern England by the sinking land and mitigated in the north by rising land.

The contrast in the climate of the British Isles is likely to become more exaggerated, for example the currently dry south-east will tend to become drier and the moist north-west will get wetter. Drought in the south-east and flooding in the north-west will both become more common. By the 2050s the climate will be about 1.5°C warmer and 8 per cent wetter than the period of 1961–90. Average sea levels will be about 35cm higher and the probability of a storm surge

If temperatures increase by 0.5°C the following may happen:	If temperatures increase by 1°C the following may happen:	If temperatures increase by 1.5°C the following may happen:
• summer and winter precipitation increases in the north-west by 2–3% • summer precipitation decreases in the south-east by 2–3% • annual run off in the south decreases 5% • frequency of the 1995 type summer (drought) increases from 1:90 to 1:25 • disappearance of a few niche species, e.g. alpine wood fern, oak fern • in-migration of some continental species and expansion of some species, e.g. Red Admiral and Painted Lady butterflies, Dartford Warbler. • increase in overall timber activity by 3% • increase in demand for irrigation water by 21% • decrease in the heating required by 6%	• summer and winter precipitation increases in the north-west by 4% • precipitation decreases in the south-east by 5% • annual run off in the south decreases by 10% • frequency of 1995 type summers (drought) increase from 1:90 to 1:10 • disappearance of certain species e.g. ptarmigan; mountain hare • expansion of the range of most butterflies, moths and birds such as Golden Eye and Redwing • increase in overall timber productivity by 7% • increase in demand for irrigation water by 42% • decrease in the heating required by 11%	• summer and winter precipitation increases in the north-west by 7% • summer precipitation decreases in the south-east by 7–8% • annual run off in the south decreases by 15% • frequency of 1995 type summer increases from 1:90 to 1:3 • further disappearance of several species • in-migration of several species • increase in overall timber productivity by 15% • increase in demand for irrigation water by 63% • decrease in the heating required by 16%

Table 11 What would be a significant climate change for the British Isles?

exceeding a given threshold will have increased. By 2050 the British Isles will be more subjected to intense rainfall events and extreme wind speeds, especially in the north. Gale frequencies will increase by about 30 per cent.

a) Soils

The soil is important in the climate change context not only because it underpins virtually all terrestrial ecosystems and agricultural production but also because it is a significant source and sink for greenhouse gases. The soils of southern Britain are likely to be most at risk from climate change, particularly in view of the predicted increase in summer droughts. This will affect the functioning of terrestrial ecosystems, influence groundwater recharge and flow and give rise to increased need for irrigation for agriculture. The combination of higher temperatures and increased rainfall may be beneficial to the soils of northern Britain and allow a wider range of crops to be grown.

Wetland areas of southern Britain will be at a risk from drying out and peat soils will be lost at increasing rates due to drying and wind erosion. Most of the clay soils of southern Britain will be subject to more intense shrinkage in the summer with potentially severe implications for building foundations.

It is predicted that soil erosion will increase throughout Britain. The predicted drier summers in southern Britain would leave the lighter soils more prone to wind erosion and wetter winters would lead to more water erosion. Higher rainfall in northern Britain would cause increased erosion by water. Over 50 per cent of grade 1 agricultural land in England and Wales lies below the 5m contour and is thus located where it might be affected by any rise in sea level.

In Britain the dominant movement of water is downwards through the soil rather than upwards via capillary rise. The implication of this is that the nutrients are leached from the soils and the soils become more acidic. The increased rainfall predicted in northern Britain will result in increased leaching of nutrients and therefore increase the acidity of soils, making them less fertile.

Organic matter is instrumental in nutrient cycling; it enhances the water holding capacity of the soil, it enhances soil stability, it improves soil structure and provides the substrate for huge numbers of soil organisms which are required for many soil processes. Climate change can affect organic matter in three main ways:

- by changing the rate in which the organic matter in the soil decays
- by altering the plant productivity rate and hence the annual return of carbon to the soil
- by altering plant species distribution and/or land use patterns and hence changing the type of plant material entering the soil.

A change from grassland to arable cultivation may reduce soil organic matter levels by 50 per cent in a period of 25 years.

There are two main considerations with respect to the impact of climate change on soil organic matter:

- the fate of organic (peat) soils

• maintenance of adequate levels of organic matter in mineral soils to maintain the fertility and stability.

In southern Britain the already endangered lowland peat, e.g. the Fens, will become even more at risk, as the combination of warmer, drier conditions in summer will cause more rapid oxidation and shrinkage. Many of the new crops that may be grown, especially in southern Britain, under a warmer climate, e.g. maize, sunflower and various legumes, are not frost tolerant and so are spring sewn. Consequently larger areas of soil than at present would have no crop cover during the winter. Increased soil erosion will result from higher winter rainfall, summer storms, the large areas of bare land and the introduction of erosion susceptible crops such as maize.

b) Flora, fauna and landscape

The natural biota of the British Isles has been most profoundly altered by human activities in the past, and over the next 50 years land use changes are likely to have an even greater impact. A 1°C increase in temperature may significantly alter the species composition in about half of the protected areas of the British Isles.

Overall the number of animal species (especially insect) in most areas is likely to increase due to immigration and expansion of species ranges. In contrast, the number of plant species may decrease and a substantial number of the 506 currently endangered species may be lost because species-rich native communities may be invaded by competitive species.

A 20–30cm increase in sea level would adversely affect mud flats and some salt marshes including nature reserves that are important for birds. Climate change will occur too rapidly for species to adapt in an evolutionary sense. Mitigating measures that can be taken include the translocation and rescue of species, the provision of habitat corridors, fire control and control of eutrophication.

There are two features about the biota of the British Isles that must be borne in mind when considering the impacts of climate change:

1. The natural landscape has been profoundly altered by human activity. The most potent forces that are active on the vegetation at present arise from the direct effects of human activity, e.g. habitat destruction by agriculture, forestry, industry, human settlements, over grazing; and indirect effects such as eutrophication through groundwater and atmospheric pollution. The overall effects of such widespread habitat disturbance will be an increase in fast growing plant species and a decrease in the slow growing stress tolerant plants of the flora, typical of unimproved grassland, lowland heath and old woodland.

2. The British Isles has a maritime climate with the small seasonal amplitude in temperature over the range that is critical for its wildlife. It spans critical latitudes at which there is a 4.5°C north/south gradient in average

summer temperatures, and sub-zero temperatures in winter. Because of these gradients a large proportion of the flora and fauna have part of their northern limits in the British Isles, often coinciding with isotherms. Hence there is a high probability that species ranges have the potential to expand northwards following climatic warming.

Cold-blooded vertebrates (the amphibia and reptiles) may become more active and noticeable during mild winters and earlier springs as was the case in 1989 and 1990, but they will not necessarily become more numerous. Their populations are already limited by the availability of habitats rather than by climate. The main effects of the high temperatures in 1984, 1990 and 1995 were an increase in the abundance, activity and geographic spread of many insects. Some of these insects were pests such as aphids in 1989 and wasps in 1990 and 1995.

The main species or communities that may become endangered by climate change include the following:

- montane/alpine and northern/arctic plant and animal species which have nowhere to go if it becomes warmer. Good examples are the tufted saxifrage and alpine woodsia fern. Animal species include the mountain hare, ptarmigan, snow bunting and white fish
- species confined to particular locations from which they cannot readily escape because, for instance, they occupy cold, damp refuges, isolated habitats or are dependent on other species for pollination, food or to complete their life cycle
- species of salt marshes and coastal communities that cannot retreat landward in the face of sea level rise.

The changes in plant communities, species, migrations, losses and gains will in time change many of the landscapes with which we are familiar:

- montane plant communities may be lost
- heaths may become subject to more frequent fires as southern Britain becomes warmer and experiences dry summers such as those in 1990 and 1995
- wetlands may dry out more frequently with a consequent change in species composition, especially if current water obstruction from aquifers is increased
- coastal dunes and rocks may be invaded more rapidly by alien species such as the hottentot fig
- salt marshes and brackish water habitats may be lost as sea level rises
- some broad leafed woodland and dry areas of Britain may decline further in response to increased frequency of summer droughts, particularly in the south where summer droughts are forecast to be more frequent and severe.

The limitations caused by reduced water availability in the south and east of the British Isles, coupled with higher temperatures and increased evapotranspiration may shift potential production of arable and other field crops northwards and westwards as well as placing extra pressure on water for irrigation. Adverse effects on soils and increased incidents of pests, weeds and diseases could reduce or negate any yield increases attributable to climatic change. Urban trees may be particularly stressed and substitute species should be sought now. Adaptation to climatic change will require positive management intervention, existing species cannot migrate and will therefore require deliberate planting.

c) Resources and industry

The high degree of sensitivity of the British water industry and water users to climate variability has been illustrated by the droughts of 1988–92 and 1995 and by the floods of 1993, 1994 and 1995. Increased winter rainfall and wetter catchment conditions are likely to increase the frequency of river flooding. It is not yet certain whether groundwater recharge will increase in future wetter winters since the recharge season could be shorter due to the increased autumn and spring evaporation.

The most important climate change impacts on building and other types of construction are likely to arise from higher summer temperatures, increased winter rainfall and, in the north of the British Isles, the increased occurrence of extreme winds. If sea defences are breached, the damage to construction in the affected areas will be considerable. Soil moisture movements on the shrinkable clays will increase, so more careful attention to foundation design will be needed in vulnerable areas. Urban traffic noise and pollution will continue to make it difficult to find acceptable solutions in town centres.

Given an overall increase in tourist activity there will be a need for better management of visitor pressure at peak periods in National Parks in order to maintain quality. Greater problems of environmental protection could arise in key settings, including crowded municipal parks, eroding beaches and moors and heaths threatened with an increased fire risk. In the north of the British Isles any significant increases in rainfall, wind speed or cloud cover are likely to offset the more general advantages associated with higher temperatures. The viability of the Scottish ski industry may decline if snow confidence becomes less secure than at present. Potential rises in sea level will adversely affect fixed waterfront facilities such as marinas and piers. On beaches backed by sea walls, increased erosion could lead to a loss of beach area.

d) The health hazards of climatic change

Direct health impacts of climate change include the following:

- deaths, illness and injury due to increased exposure to heatwaves
- reduction in cold related diseases/disorders

- altered rates of death, illness and injury due to changes in frequency or intensity of climate related disasters (droughts, floods, forest fires, etc.).

Indirect impacts include:

- altered distribution and transmission of vector borne infectious diseases (viral infections, malaria, etc.)
- altered distribution and transmission of certain communicable diseases (water borne and food borne infections, some respiratory infections, etc.)
- impacts on agriculture and other food production; beneficial effects may occur in some temperate zones
- the effects on respiratory systems via increased exposure to pollens, spores and certain air pollutants
- consequences of sea level rise via flooding, disrupted sanitation, soil and water salination and altered breeding sites for infectious disease vectors
- impacts on health caused by demographic disruption, displacement and a decline in socioeconomic circumstances as might result from impacts of climatic change on natural and managed ecosystems.

Recent unexpected outbreaks of infectious diseases such as cholera, Dengue Fever and Hantavirus are reminders that surprises occur frequently with infectious diseases. We cannot expect to foresee clearly the impacts of climate change on all infections. Malaria is an important example with around 350 million new cases worldwide annually, including approximately 2 million deaths. Malaria existed in the British Isles for several thousand years and it is only within the past half century that it has been effectively eradicated, largely through drainage of wetlands and the use of insecticides, especially in southeast England. With changing climate there is increased potential for the transmission of malaria in much of Europe but the existing environment and public health defences should be sufficient to prevent its reintroduction. Increased mean summer and winter temperatures and wetter winters will affect the proliferation of arthropod vectors (e.g. ticks or mites), since these are more sensitive to temperature than the availability of food. Sea level rise may increase flooding and the presence of brackish water which may also encourage some insect vector species.

There are seasonal patterns in the incidence of food borne illness in the British Isles. Food borne illness in England and Wales during 1982–91 occurred much more frequently in the late summer months, and a particularly strong relationship was observed between the incidence of food borne illness and temperature in the preceding month, suggesting that the high ambient temperatures have their most significant impact at the point in the food system prior to the food reaching the consumer.

Conditions which favour the occurrence of blooms of blue/green algae include hot summers and nutrient rich waters (caused by changes in agricultural run off or reduced river flow) and therefore

Positive effects	Negative effects
• an increase in timber yields (up to 25 per cent by 2050) especially the north of Britain • a northwards shift of farming zones by about 200–300km per degree centigrade of warming, or 50–80km per decade. This will improve some forms of agriculture especially pastoral farming in the north-western part of Britain • enhanced potential for tourism and recreation as a result of increased temperatures and reduced precipitation in the summer, especially in southern Britain.	• an increase in the drought, soil erosion and the shrinkage of clay soils • an increase in animal, especially insect, species as a result of northwards migration from Europe and a small decrease in the number of plant species due to the loss of northern and montane species • a decrease in crop yields in the south-east of Britain • an increase in river flow in the winter and a decrease in the summer, especially in the south • an increase in public and agricultural demand for water • increased damage effects of increased storminess, flooding and erosion on natural and human resources and human resource assets in coastal areas • increased incidents of certain infectious diseases in humans and of the health effects of episodes of extreme temperature.

Table 12 The likely effects of a changing climate in the British Isles

the incidence of freshwater toxic algal blooms may increase with climate change.

4 Ozone depletion

The atmosphere of the Earth consists of about 78 per cent nitrogen, 21 per cent oxygen, 0.4 per cent carbon dioxide and 1.3 per cent argon. The amount of ozone in the atmosphere is a small but vital component of the atmospheric composition. Ozone occurs because oxygen rising up from the top of the troposphere reacts under the influence of sunlight to form ozone. Most of this is created over the equator and the tropics because this is where solar radiation is strongest. However, winds within the stratosphere transport the ozone towards the polar regions where it tends to concentrate.

Ozone is constantly being produced and destroyed in the stratosphere and maintains a natural dynamic balance. As well as

Figure 34 Computer image to show ozone hole over Antarctica

being produced by sunlight it is also being destroyed by nitrogen oxides. However, although ozone is constantly produced and destroyed, human activities may tilt the balance one way or the other. There is now clear evidence that human activities have led to the creation of a hole in the ozone layer over Antarctica (Figure 34).

The hole in the ozone layer over Antarctica was first discovered in 1982. It follows a very clear seasonal pattern; each spring time in Antarctica (between September and October) there is a huge reduction in the amount of ozone in the stratosphere, but as the summer develops the concentration of ozone recovers. What causes the depletion in ozone during the spring time?

During winter in the southern hemisphere the air over Antarctica is cut off from the rest of the atmosphere by circumpolar winds; these winds block warm air from entering into Antarctica. Hence the temperature over Antarctica becomes very cold, often as low as $-90°C$ in the stratosphere and this allows the formation of clouds formed of ice particles. Chemical reactions take place on this ice which involve chlorine compounds resulting from pollution by human activities. These reactions release chlorine atoms. Once the Sun returns during the spring the chlorine releases atomic chlorine which destroys ozone in a series of chemical reactions. Hence the hole in the ozone layer occurs very rapidly in the spring. By summer, however, the ice clouds have evaporated and the chlorine is converted to other compounds such as chlorine nitrate until the following winter. The ozone hole fills in, although it returns each spring. The size of the ozone hole is also huge (Figure 34); in 1987 it covered an area the size of continental USA and was as deep as Mount Everest.

Nitrogen oxides produced by human activities can also destroy the ozone layer just as chlorine atoms can. The sources of chlorine atoms are known as Chlorofluorocarbons (CFCs) and include materials used in fridges, air cooling systems, foamed plastic and aerosols. CFCs are particularly dangerous because they can be very long lived (over 100 years) and they spread throughout the atmosphere. In the case of Antarctica the build up of chlorine appears to have had very little impact until it reached a critical threshold. Once that threshold was reached only a small increase in chlorine led to a huge change in the ozone layer.

There are major implications of an increase in the hole in the ozone layer as it means that ultraviolet radiation will reach the ground in increased quantities. Some ultraviolet reaches the ground already, it is in the 290–320 wave band. This is known as UV-B, and can cause sun burn, skin cancer and eye problems such as cataracts. It is estimated that for every 1 per cent decrease in the concentration of ozone there will be a 5 per cent increase in the amount of skin cancers each year. Crops and animals have also been tested to see how they react with an increase in UV-B radiation; soya bean experiences a 25 per cent decline in yield when UV-B increases by 25 per cent, while cattle are affected by eye complaints including cancer of the eye.

5 Policies to combat climate change

Emissions of the main anthropogenic greenhouse gas, CO_2 are influenced by

- the size of the human population
- the amount of energy used per person
- the level of emissions resulting from that use of energy.

Similar factors affect the levels of emissions of the other greenhouse gases.

A variety of technical options could reduce emissions, especially from use of energy. Reducing CO_2 emissions can be achieved through:

- improved energy efficiency
- fuel switching
- use of renewable energy sources
- nuclear power
- capture and storage of CO_2
- increasing the rate at which natural sinks take up CO_2 from the atmosphere, e.g. by increasing the amount of forests.

Reductions in emissions of other greenhouse gases can also be achieved using technology. Methods of reducing methane emissions include reductions in leakage and capture of emissions, with destruction or utilisation of the methane.

a) International policies to protect climate

The 1985 Vienna Convention and the 1987 Montreal Protocol on substances that deplete the ozone layer are major examples of international agreements to protect the atmosphere. The 1987 Montreal Protocol was the crucial first step in limiting further damage to the ozone layer in the stratosphere. It was signed by many countries to greatly reduce the production and use of CFCs which had been shown to be responsible for causing the damage. Since 1987, further amendments to the protocol have imposed even greater restrictions of the production and use of potentially damaging compounds. Two revisions of this agreement have been made in the light of advances in scientific understanding, the latest being in 1992. Recognising their responsibility to the global environment the countries of the EU agreed to halt production of the main CFCs from the beginning of 1995. It was anticipated that these limitations would lead to a recovery of the ozone layer within 50 years, but recent investigations suggest the problem is perhaps on a much larger scale than anticipated.

The Toronto conference of 1988 called for the reduction of CO_2 emissions by 20 per cent of the 1988 levels by 2005. Also in 1988 the Intergovernmental Panel on Climate Change (IPCC) was established by UNEP and the World Meteorological Organisation. The UN Conference on the Environment and Development (UNCED) was held in 1992 in Rio de Janeiro. It covered a range of subjects and there were a number of statements including the Framework Convention on Climate Change (FCCC). This came into force in March 1994.

Joint implementation (JI) was outlined in the FCCC. The theory of JI is that maximum reduction in emissions of greenhouse gases can be achieved at lowest cost through international co-operation. Thus, JI is a mechanism by which countries may reach their target of CO_2 stabilisation by reducing greenhouse gas emissions at home and by working with other countries, where emission reductions may be achieved at lower cost.

Countries active in AIJ (*activities implemented jointly*) include Australia, Canada, Japan, Netherlands, Norway, Sweden and the USA. A range of AIJ projects are now operating including forestry schemes, fuel switching, wind power, hydro-electric power, geothermal power and landfill gas. The main drawbacks to AIJ are that they impose additional costs on international projects due to the monitoring and verification needed to ensure that the expected reduction in emissions is achieved in practice. Nevertheless, this is a rapidly growing area of international co-operation.

b) Reducing emissions

The amounts of greenhouse gases in the atmosphere can be reduced either by controlling emissions or by increasing the rate at which they are removed. The main anthropogenic sources of CO_2 are combustion of fossil fuels, cement manufacture and deforestation. The second most important greenhouse gas is methane. Anthropogenic sources of methane emit about 375 million tonnes per year (compared with natural sources which produce only about 160 million tonnes per year). Methane is emitted as a result of production and use of fossil fuels but much more comes from disposal of solid and liquid wastes and agriculture (especially the growth of rice and from ruminant livestock, such as cattle).

Reducing emissions of CO_2 could be achieved by:

- improving energy efficiency
- switching to low carbon fuels
- switching to 'no-carbon' fuels.

The first two options are cost effective, are used in many places and deliver useful reductions; whether this will be sufficient is not yet known. Large reductions could be provided by the third option, such as switching to renewable energy or nuclear power, but the world is heavily dependent on fossil fuels. Therefore, it is important to use technology options which will allow continued use of fossil fuels without substantial emissions of CO_2; this can be achieved with capture and storage of CO_2 from flue gases. Increasing the rate of removal of CO_2 from the atmosphere through enhancing natural sinks for CO_2, such as growth of forests or biomass in the oceans, could also contribute to reducing atmospheric concentrations of greenhouse gases.

Similarly a range of measures could be employed to tackle methane emissions:

- production of methane can be limited
- leakage can be controlled
- emissions can be captured and put to use
- methane can be destroyed.

In 1992 at the Earth Summit in Rio de Janeiro, the United Nations Framework Convention on Climate Change was signed. This expressed concern about the effects of climate change. In 1997, governments established the Kyoto protocol, ratification of which would set up targets for reduction of greenhouse gas emissions.

MEDCs agreed to cut their collective greenhouse emissions by an average of 5.2 per cent below 1990 levels over the years 2008–12. Different countries adopted different targets: the EU committed to a cut of 8 per cent, the USA to 7 per cent and Japan to 6 per cent. Russia and the Ukraine agreed to stabilise at 1990 levels.

The Hague Conference in 2000 was described as a make-or-break gathering at which countries accepted or refused the terms of the Kyoto protocol. In order for the protocol to come into practice, it had to be ratified by 55 parties. Some of them had to be developed industrial countries, which account for 55 per cent of the carbon dioxide emissions. However, there were problems with the Kyoto protocol; many pressure groups were concerned about a technical detail in the protocol, which allows a participating country meeting its emissions reduction target to trade any surplus reduction with another country. This means richer countries could 'buy' such reductions and avoid making significant cuts to their emissions. The USA, Canada, Japan, New Zealand and Australia want most of their emission targets to be fulfilled by these emissions-trading mechanisms. The EU has been keener to make real cuts in emissions, but many EU governments, including Britain, want the USA to ratify the Kyoto agreement, so may be forced to make concessions. Meanwhile, the LEDCs are divided on some issues and in agreement on others. The key forces are China and India. Saudi Arabia and other OPEC countries are asking for compensation for lost revenues from the export of oil.

Some countries have found it easier to make the cuts stipulated by the Kyoto protocol than others. Britain, for example, had already reduced its use of coal and oil and changed to natural gas. Germany, too, had exhausted many of its traditional heavy industries, such as iron and steel which consume vast quantities of coal, and Japan was largely dependent on nuclear power and hydro-electric power. In contrast, the USA was experiencing an economic boom in the 1990s and was reluctant to make cuts.

There are three main ways for countries to keep to the Kyoto target without cutting domestic emissions:

- plant forests to absorb CO_2 or change agricultural practices e.g. keeping fewer cattle
- install clean technology in other countries and claim CO_2 reduction credits for themselves
- buy carbon credits from countries such as Russia where traditional heavy industries have declined and the national CO_2 limits are underused.

Even if greenhouse gas production is cut by 60–80 per cent there is still enough greenhouse gas in the atmosphere to raise temperatures by 5°C. The Kyoto agreement was only meant to be the beginning of a long-term process, not the end of one. The guidelines for measuring and cutting greenhouse gases were not finished in Kyoto. It was not decided to what extent the planting of forests and CO_2 trading could be relied upon. Since George W. Bush was elected President of the USA he has rejected the Kyoto protocol on the grounds that it would hurt the US economy and employment. Although the rest of the world could proceed without the USA it emits about 25 per cent of the world's greenhouse gases. So without the USA, and LEDCs such as China and India, the reduction of carbon emission would be seriously hampered. According to the Kyoto rules, 55 countries must ratify the agreement to make it legally binding worldwide, and 55 per cent of the emissions must come from MEDCs. If the EU, eastern Europe, Japan and Russia agree, they could just make up 55 per cent of the MEDCs' emissions. Without the USA (and Australia and Canada who are against cutting emissions) it is difficult to achieve this goal.

Without agreement, the world's poor will suffer. Flooding, drought and famine will continue. As we have seen, the losses appear to outweigh the gains. Productive land will be submerged; there will be more natural disasters; a collapse of the insurance industry is possible, and this could lead to a global crash. However, if we can cut fuel use, emissions of greenhouse gases, and use more renewable sources of energy, conditions might not deteriorate as much.

Summary

- Climate change appears to be accelerating. Much of it is natural, but increasingly more of it is due to human activities.
- More and more people are at risk from climate change, not just in LEDCs but in MEDCs such as the USA.
- Many of the areas most 'at risk' of climate change are densely populated urban areas.
- The correlation between increased greenhouse gases and global warming is well established, but natural changes have an impact too.
- In the British Isles the impacts of climate change would affect soils, flora, fauna, agriculture, water resources, tourism and human health.
- Human activity is also affecting the ozone layer. Depletion of the ozone layer is linked to a rise in skin cancer, eye problems, and affects plant yields.
- There have been many attempts to tackle climate problems. These date from the 1970s but have had little success so far. One of the stumbling points appears to be whether countries fear economic loss over environmental gain. If countries feel they will lose out economically they will not protect the environment.

Questions

1. Refer to Figure 35.
   a) Describe the pattern of temperature change.
   b) Comment on the possible impacts that this will have.
   c) Describe the form of the two graphs and comment on their relationship.
   d) Describe and comment on the emission of CO_2 from fossil fuel combustion.
2. With the use of examples, explain why attempts to manage climate change have had limited success.

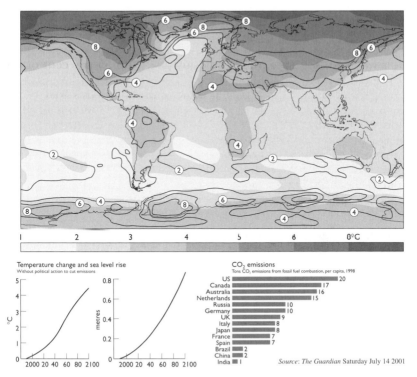

Figure 35 The annual mean change of the temperature and its range
between now and 2100 if the world does not act on warnings

Advice

I. a) Temperature changes increase towards the poles (I mark) especially
the North Pole (I) and are greater over continents compared with
oceans (I). Changes in the northern hemisphere are greater than in
the southern hemisphere (I) and the lowest changes (less than 2°C)
are the tropical and sub-tropical areas of the southern hemisphere (I).

b) A wide range including rising sea levels; increased storm frequency
and magnitude; melting ice caps; increased rainfall and increased
drought; impacts on farming, tourism, etc., water metering, hosepipe
bans, etc.

c) Both are positive (I); and follow a similar trend (I); temperatures
increase fairly steadily (with a very slight j-shape) (I) to reach an
increase of 4.5°C over 110 years (I). Sea level rises slightly more
steadily by 0.8m (I) over 110 years.

d) The largest emitters agree MEDCs – e.g. the USA $20t/CO_2$ per
capita, Canada $17t/CO_2$ and Australia $16t/CO_2$. The top 11 countries
are MEDCs – the EU, North America, Japan, CIS and very large

countries such as India, China and Brazil. These countries use much coal in order to industrialise.

2. Apparently straightforward, it is still possible to argue that policies have been effective or at least have had some impact.
This is an 'explain why' question, and will require discussion on factors such as

- resource availability
- state of the economy
- type of economy
- power of the energy sector
- stage of development
- problems of cooperation.

Candidates should include in their discussion some details regarding the Montreal Protocol, the Kyoto Conference and the Hague/Bonn meetings. The most likely case studies will be

- US energy sector (the Bush administration) versus a Kyoto/Bonn
- LEDCs desiring to develop their economies
- MEDCs such as the USA wishing to maintain their economic strength
- countries which have the potential (resource and capital) to develop alternative energy sources.

Further Reading

R. Barry and R. Chorley, *Atmosphere, Weather and Climate* (Routledge, 1998)
D. Elsom, *Smog Alert: managing urban air quality* (Earthscan, 1996)
A.S. Goudie, *Environmental Change* (Oxford, 1992)
J. Gribbin, *The Breathing Planet* (Blackwell and New Scientist, 1986)
M. Hulme and E. Barrow, *Climates of the British Isles* (Routledge, 1997)
H. Lamb, *Climate, History and the Modern World* (Methuen, 1982)
E. Linacre and B. Geerts, *Climates and Weather Explained* (Routledge, 1997)
G. Nagle, *Hazards* (Nelson, 1999)
G. Nagle, 2000 *Britain's Changing Environment* (Nelson 2000)
R. Rotberg and T. Rabb, *Climate and History* (Princeton, 1981)

Web sites
http:www.bbc.co.uk/weather/ is the BBC Weather Centre with scientific information, weather lore, forecasts and so on
http:www.meto.govt.uk/ is the Met Office Home Page and contains a large amount of educational and forecasting products.
http:witu.rdg.ac.uk/rms/rms/html is the Home Page of the Royal Meteorological Society and has excellent links to weather and climate in the UK and worldwide
http://www.met.reading.ac.uk/~swstheri/WAMP1.html for the West African Monsoon project.
www.nelsonthornes.com for a monthly update of weather news and hazards

Index